Advance Praise for

NERVOUS

"The essays in *Nervous* crackle and pulse with a beautiful bodily wisdom that animates a sparkling intellect. Jen Soriano tenderly, unflinchingly excavates layers of history and pain—found both in her body and our body politic—and offers all of us tools and materials to build a path toward wholeness. I'm in awe of Jen Soriano, and you will be, too."

—Angela Garbes, author of *Essential Labor* and *Like a Mother*

"As I neared the end of this viscerally moving book, I thought of my students hungrily absorbing the stories in *Nervous* along with the stories in Maxine Hong Kingston's *The Woman Warrior* and Cathy Park Hong's *Minor Feelings*. . . . *Nervous* has instantly joined the crucial works of Asian American literature and the newly teeming space of American literature as a whole. A brilliant reckoning, *Nervous* begins in Soriano's individual story but enlarges to include the necessary stories of family, community, and homelands. 'When we are met with erasure, we story back with permanence,' Soriano says. Though it is steeped in pain, *Nervous* is nevertheless a testament of exultant embodiment—of woundedness and remedy, of memory and history, of disruption and coalition, of diaspora and belonging."

—Rick Barot, author of *The Galleons* and *Chord*

"*Nervous* is the epitome of innovation. In this painfully glorious essay collection, Jen Soriano illuminates the ways Filipinos have been mistreated and oppressed by a multitude of systems, both in colonial times and in the present. That is the best of what a collection can do: giving voice to the silent corners we've been forced into. *Nervous* accomplishes that and much more—a true literary achievement."

—Evette Dionne, award-winning author of *Lifting as We Climb* and *Weightless*

"This book is such a gift! Part medical history, part lyrical memoir, Jen Soriano traces the rivers and tributaries of her pain, becoming fluent in the language of her body. *Nervous* is a revelation for every person who has been silenced, neglected, and made to feel unworthy of care. Luminous and tender, *Nervous* is not your conventional trauma narrative."

—Alice Wong, founder and director of the Disability Visibility Project and author of *Year of the Tiger*

"I couldn't put it down. This book brings light to the dark tunnels of history that live in our bodies. I wish for all mental health, social services, and wellness practitioners to read *Nervous*."

—Dr. Leny Mendoza Strobel, author of *Coming Full Circle* and editor of *Babaylan: Filipinos and the Call of the Indigenous*

"I often wonder how many paths there are when writing about a mental health journey. Sometimes writers lean toward the facts—statistics, historical contexts, and slightly more creative retellings of academic texts. Other times, writers choose the personal: narratives of lives lived, tears cried, hearts broken, illnesses ignored or dismissed. In *Nervous*, Jen Soriano has found the most intriguing and captivating way to do both, which helps the reader understand how the historical dismissal of women's pain directly impacts how we are treated today. By showing us how real she is, Soriano forces us to consider the humanity of what pain they hid, who they hid it from, and what their lives could have been if they had been permitted to feel anything. *Nervous* takes the focus from the abstract and does what doctors (and historians) failed to do: makes her story, her pain, and her life as real as any history that preceded. *Nervous* gives face and weight to those forgotten women whose suffering has become little more than anecdotal collections of stories, not real people. It's seamless and powerful. *Nervous* is a masterful personal narrative, beautifully written and captivating. It should—and will—be placed alongside some of the best well-crafted and compelling contemporary memoirs of this era."

—Bassey Ikpi, *New York Times* bestselling author of *I'm Telling the Truth, but I'm Lying*

NERVOUS

NERVOUS

ESSAYS ON HERITAGE AND HEALING

JEN SORIANO

AMISTAD

An Imprint of HarperCollins*Publishers*

This book should not be considered a substitute for medical treatment, psychotherapy, or advice from a mental health professional.

Grateful acknowledgment is made to the following for permission to reprint excerpts of previously published material:

Excerpt on page 3 from the book *Hard Times Require Furious Dancing*. Copyright © 2010 by Alice Walker. Reprinted with permission of New World Library, Novato, CA. www.newworldlibrary.com.

Excerpt on page 19 from "On Migration, Upon Finding an Old Map" by Michelle Peñaloza, in *Former Possessions of the Spanish Empire*, Inlandia Institute, © 2019. Reprinted with permission of the author.

Excerpt on page 85 from "State of Siege" by Eric Gamalinda, in *Lyrics from a Dead Language*, Anvil Publishing, © 1991. Reprinted with permission of the author.

Excerpt on page 187 from "Becoming a Minke Whale" by Angela Peñaredondo, in *nature felt but never apprehended*, Noemi Press, © 2023. Reprinted with permission of the author.

Excerpt on page 221 from "Anodyne" from *Pleasure Dome: New and Collected Poems* ©2001 by Yusef Komunyakaa. Published by Wesleyan University Press. Reprinted with permission.

Excerpt on page 237 from "mostly water" by Arlene Biala, in *one inch punch*, Word Poetry, © 2019. Reprinted with permission of the author.

"A Brief History of Her Pain" was originally published in *Waxwing*, Issue X, Fall 2016. Reprinted with permission.

A previous version of "War-Fire" was originally published in *Fugue 57*, Summer/Fall 2019. Reprinted with permission.

Thanks to ARTS By The People for publishing earlier versions of "A Brief History of Her Pain," "War-Fire," and "Unbroken Water" in the chapbook *Making the Tongue Dry*, © 2019.

FIRST EDITION

Designed by THE COSMIC LION

Illustrations on pages ix, 19, 20, 85, 86, 133, 134, 187, 188, 221, and 222 © Modulo 18/Shutterstock

Library of Congress Cataloging-in-Publication Data has been applied for.

ISBN 978-0-06-323013-2

23 24 25 26 27 LBC 5 4 3 2 1

To the silenced.
May your words flow free.

Historical trauma is a story of love.

—KARINA WALTERS

What we have is a knowing body.
I want to pit it against the knowledge
hierarchy of Western modernity.

—MERLINDA BOBIS

Our nervous systems are trying
to do the right thing.

—STEPHEN PORGES

Contents

IV
Neuroplasticity

V
Neuromimicry

Author's Note

I have thoroughly researched the information in *Nervous,* but I am not a neuroscientist or a physician, and this book should not be considered a substitute for medical treatment, psychotherapy, or advice from a mental health professional. If you are experiencing depression, abuse, addiction, suicidality, or other severe emotional distress, please seek professional help immediately. If you are hesitant to seek help, I hope this book might inspire you to reconsider.

This is a work of nonfiction, but it is also a work of memory. My memories of events are bound to be different from those of the people mentioned in the essays. I have rendered these events as faithfully as possible to my own memory and in some cases have checked details with others who were there. Though conversations come from my keen recollection of them, they are not written to represent word-for-word documentation; rather, I've retold them to evoke the emotional truth of what was said, in keeping with the essence of the moments.

A note on terminology: Language is imperfect and fluid, and therein lies its beauty. There is no official rule for how to name Filipina/o/x Americans. And if there were, I would probably defy it, as we all might, in order to better reflect the diversity in our millions. I have chosen to use the term Filipino when describing Filipino communities and populations, both in the United States and in the Philippines, not because I believe it to be gender-neutral, as some claim, but because it is the most recognized umbrella term for our people. I also believe that Filipino is a more accurate term to apply to the big tent of our diaspora than is Filipinx. There is a queer and radical politic inherent to the term Filipinx that can be preserved for those who identify with its origins and (evolving) meanings.

Beyond using Filipino as a big tent term, I apply its variations with as much specificity as possible; I use Filipina when referring to women and femmes, Filipino when referring to men and masculine of center people, and Filipinx when referring to nonbinary, genderqueer, and gender fluid people. I also use Filipinx to describe groups who claim the term for political reasons I have alluded to above. These reasons include but are not limited to centering LGBTQIA+ people, acknowledging the gender spectrum, and aligning in solidarity with Latinx and Latine communities.

The following names are pseudonyms: Mark, Susan X, Kay, Marina, and Violet.

Last, before you read, you might want to know that *Nervous* is not a traditional trauma narrative, but there are traumatic events described in the essays. "381 Years" contains primary source accounts of violence from war and colonization in the Philippines, "Bayanihan" tells stories of suicide attempts and suicidal ideation, and "Broken Water" recounts violent delusions about pregnancy and birth.

The point of this collection is to allow pain to surface, but also to swim with strong currents of love and humor and joy.

Introduction

We are nervous beings, in nervous nations, at an increasingly nervous time. I wrote this book for those of us who have felt crazy and alone. For those of us who have been told to forget about the past, to stop being weak, and to swallow our pain. For all of us with a knowing body. For babaylan who remember a time before silence. For pearls in their shells seeking conditions to shine. My story is just one ripple in an emerging ecosystem of interdependence, where we don't have to bear generational pain alone.

~

This is a story I'm not supposed to tell. I'm supposed to continue a lineage of silence—be wordless in pain, resilient and productive, a walking American dream. But my body speaks in a language of grief and suffering and with words of longing, liberation, and love.

I have always been a nervous person. I startle at the sound of my name. I enter rooms searching for exits. I worry about being invisible while constantly afraid of becoming seen.

Perhaps because I have been so nervous, I am also a person in pain. My hands smolder and my wrists feel riddled with broken glass. On any given day, my nerves are either drowned or lit on fire.

Nerves might make you think of fainting rooms, smelling salts, and frail maidens fanning pale faces.

Nerves make me think of machetes and murder, survival and silence, the sibilance of stories untold.

~

For a long time, I didn't talk about my pain because I wanted to hide it, like you might hide soiled panties or some other mortifying mistake. It wasn't until I read Audre Lorde's *The Cancer Journals* that it occurred to me I didn't have to be so silent and afraid.

~

I started writing *Nervous* after being diagnosed with central sensitivity syndrome, an umbrella term for various conditions affecting the nervous system, including fibromyalgia, irritable bowel syndrome, and chronic pain from central sensitization. I was a new parent learning to raise my first and only child while also embarking on a journey to understand the pain I'd carried for most of my life.

Years before this I was also diagnosed with peripheral neuralgia, dystonia, mild scoliosis, generalized anxiety disorder, social anxiety disorder, and complex post-traumatic stress disorder (C-PTSD). I had found these diagnoses affirming, but I didn't yet know what lay at the root. What traumas had disrupted my nervous system to yield such pain and to give rise to these comorbid conditions? And how could I keep my child from a similar fate?

This collection traces my journey of asking questions and finding answers, breaking silence, and becoming seen.

~

Though pain and trauma are highly individualized experiences, and by definition can make us feel cut off from the rest of the world, they are the reality for the majority of us. Seven out of ten of us go through at least one traumatic event in our lifetimes. About one in ten of us will go on to develop PTSD or C-PTSD. Some professionals consider post-traumatic impacts to be a profound public health problem, with as many as one in five people worldwide (and twelve million people in the US alone) experiencing PTSD in any given year.

The fact that psychological trauma is so widespread makes it a collective rather than a solely individual experience. Similar to how the COVID-19 pandemic affected the entire globe, trauma affects a multitude of bodies and the body politic at large. This collective experience is not just about suffering. Grief and pain may be carried in our nerves, but so are chronic joy, laughter, and the ancestral wisdom that comes from our somatic impulse to survive.

A Brief History of Her Pain

I will keep
Broken
Things.

I will keep
You:

Pilgrim
Of
Sorrow.

I will keep
Myself.

—ALICE WALKER

AD 1996

Jennifer S. Asian. Nineteen years of age. Clinic visit due to mysterious pain "like lead shooting through arms." Average build, height. Flat affect. Patient enters clinic clenching then twisting her fists, as if juicing two lemons till dry.

Jennifer S. received standard procedure for unexplained pain:

- The Wartenberg Wheel: rolled steel teeth across inner forearms and palms to assess nerve response.

- The Tomahawk Hammer: struck hammer against wrists, knees, elbows to elicit deep tendon reflexes.

Results: No nerve damage. No delay in deep tendon reflexes. Pain may result from emotional overreaction to college stress.

Jennifer S. sent home with naproxen, medical report, reminder to schedule a pap smear.

1900 BC

Egyptian healers write the Kahun Papyrus, the first comprehensive medical document known to man, which happens to be focused entirely on women's health. The Papyrus mentions a condition of mysterious pain and mercurial emotional states. This is what will later become known as hysteria. The text also reveals that, though seemingly mysterious, the condition does indeed have a physical source, and it is none other than a dislocated or starved uterus.

To treat the condition, sweet oils must be applied to the vagina, or unpleasant things must be consumed to lure the misbehaving womb back into place.

The Kahun Papyrus may not only be the first medical text known to man; it may also be the first text written by women. The authors may have been priestesses, since healing was within the domain of spiritual leaders. But being women, these priestesses may have had misplaced wombs, and therefore may have also been physically pained, emotionally disordered, and hysterical.

AD 2001

To this day I wonder what happened to her—the priestess-like bruha in yoga pants. I don't remember her name, but she had brown hair, long like her willow-branch arms, and the slinky bearings of a catwalk model.

I find her at a neighborhood healing studio. I've walked in because I'm desperate for pain relief. Toward the end of the massage, she asks if she can share something; she had heard a word emanating from my left shoulder.

"It's a word I never use," she says. "I don't even know what it means."

"What is it?" I ask.

"Remorse," she says. "Does that mean anything to you?"

"Remorse," I repeat, feeling a slight chill of fear on my skin. "Yes, actually, it does."

"Well, whatever it means, remorse runs through your shoulder."

Remorse runs through my shoulder. What a poetic way to describe the draconian pain that has seized that piece of my body for years.

I am stunned and a bit creeped out. How can this willowy woman hear the reverberations of my pain and interpret them to the point of uttering a word she doesn't understand? Then, gratitude begins to glow in my chest. She has just affirmed my pain in a way that no doctor ever has. *Maybe I'm not crazy after all,* I think. *Or at least, maybe I'm not alone in my madness.*

Later, when I've turned over onto my back, she taps my forehead with her pointer finger, then moves away from the table with her head down, an intense look of focus on her face.

"There were two purple orbs on your right side, and now they are surrounding your whole body," she says, still looking down. "I'm supposed to tell you that they are enough. They are enough to support you."

As she pronounces this, my eyes grow wet with tears. It's a reflexive reaction—a somatic response to feeling seen. Support and protection were nice ideas. As in, they seemed to be nice things to have. Some of my body pain, I had come to believe, arose not just from external stress but from living in a perpetual state of bracing in self-support, for want of external protection.

I want to say this to her. I want to affirm whatever and whomever she is channeling. Instead, I simply ask, "Do you always see things when you give massages?"

"Not always." She returns to the table, her hands hovering inches above my left shoulder. "But it's something that runs in my family. My mother used to say, 'Get the phone,' and five seconds later, the phone would ring."

In retrospect, I wonder if her witchy powers forewarned her of what was to come. A few weeks later, I arrive at our next session eager to ask more about the purple orbs and remorseful shoulder. But she never arrives. On the way to the studio, her bike hit a car, or a car hit her bike, I'm not sure which. I just know she had an accident, which she supposedly walked away from, but I never saw her again.

500 BC

Hippocrates and his followers affirm the Kahun Papyrus by diagnosing emotionally erratic behavior as a disease caused by a wandering womb. Hippocratic doctors one-up the authors of the Papyrus by claiming that hysterics' wombs are accident-prone: they have a tendency to move upward and then fall, causing severe bodily pain wherever the wombs may land.

They explain that in some cases, the uterus can attach itself to the bladder, which then prevents sperm from entering, causing the uterus to rot in that spot and wither away. In a better-case scenario, the uterus can move upward and crowd other organs, resulting in, as one might imagine, all sorts of hysterical problems.

AD 2002

Jennifer S., twenty-six years old, Philippina. Has two older brothers. First time in therapy. Presents as anxious. Emotional. Used entire box of tissue. Complained of pain in arms, back, neck. Called it "debilitating." Asked her about childhood. Silence. Mentioned silence multiple times.

Offered her a framework: taboos in Asian culture. Jennifer S. became animated, articulate: "I didn't pay for an anthropologist. Plus, you're wrong." Anger issues. Probed possibility of sexual abuse. Patient claimed relationship with brothers was positive: "One of the best parts of childhood." Defensive affect. Highly strung. Patient left crying. Gave her article on breathing techniques for pain.

AD 200

The Roman physician Galen agrees that the womb causes problems but disagrees on the nature of its errors. Hysteria is not due to the uterus's itinerant movement, he says, but to its unfortunate ability to retain blood and also, of course, female sperm.

The treatment? Not so unfortunate, and in that era, not yet taboo: pelvic massage, sexual stimulation, and release—better known as orgasm. So begins a millennium-long heyday of treating hysteria through sexual healing.

AD 2002

I am sitting on an examination table, swinging my legs to distract myself from pain while yet another doctor asks the requisite questions.

"How long have you had this pain?" The doctor looks at me squinty-eyed, like she is having trouble seeing me, though I am only two feet away.

"More than ten years," I reply.

"What caused it?"

"I don't know. I'm hoping you can tell me."

"What makes it better?"

"Muscle relaxers. Moderate exercise. Sex," I say.

The doctor looks at me sharply and, I think, disapprovingly. I have overestimated the amount of camaraderie we would have as two relatively young women of color.

"I've read about JFK and back pain," I prattle on. "And about how he was addicted to sex because dopamine supposedly helped ease the pain."

Her head jerks up and she asks, "Have you ever been on opioids?"

"No. Other doctors say they wouldn't help my pain, but sometimes I wonder."

"Hm," she nods. She's probably relieved I'm not an opioid addict.

"Oh, and I have a sharp burning pain in my abdomen and have to pee all the time," I recite. I've rehearsed this line with a doctor friend who told me these magic words will get me an MRI, one of the few diagnostic tests I have not done.

The doctor looks sharply at me for a third time, her eyes nearly pinched closed, as if she's hoping I'll disappear. I wonder if this is a tic of hers and if she might be a bit hysterical herself.

She reaches over to palpate my belly. Perhaps she's checking to see if my uterus is in place.

"Ow," I remember to say after a brief pause.

My performance is unconvincing. She sends me home with a diagnosis of "unexplained chronic pain" and a bottle of Trazodone—an antidepressant and mild sleep aid. To me, the text on the bottle screams, "Go to sleep and shut up!" I'm disappointed but not surprised. I've become used to doctors treating my pain like an imaginary friend.

I go home to my boyfriend, Mark, and pop a Trazodone like a Tic Tac. The effect is pleasant. Mark's chest bends and blurs above me as we dance through a slow, sweet, sweaty session of sex that stretches out like saltwater taffy. The dopamine rush of orgasm temporarily washes away my pain. Until the crash. Then, like an addict, I count the hours until my next fix.

Mark and I have had sex nearly every night since we started dating. For the past two years, sex has been our chosen form of communication. Who wants to plumb the depths of feelings when, uncaged, they could eat you alive? Better to pet what's on the surface, where it's safer and much more fun.

Together Mark and I move like cats, independent and tactile and self-serving. We palm limestone walls and soak in sulfured hot-spring pools. We fall asleep with limbs intertwined and wake stroking hairs on naked forearms and calves. This sensuous and silent way of being is a necessary balm to my pain.

AD 1350

The rise of Christianity and Catholicism in Europe brings a more—dare we say—holistic diagnosis of hysteria: a misbehaved womb is just one part of a whole, unholy, transgressive woman.

Dominican friar St. Thomas Aquinas marries hysteria with evil, and the sick woman witch is born. He concludes that because of weakness, women are more prone to temptation from the Devil, one result of which is the curse of hysteria. To top it off, St. Thomas declares that all unstudied women who practice healing of any kind are minions of the Devil himself.

The punishment is cure, and the cure is punishment: exorcism.

AD 2003

After two years of sex, camping trips, sex, hot-spring soaking, sex, music sharing, sex, and friendly passing of days, Mark and I break up because I've been cheating.

My real relationship is with the smoldering pain that possesses me. It is this relationship that receives all my attention, the majority of my money, the totality of the scarce nurturing I have to give.

I am engulfed in a war between nerve signals that say I am hurt but will be fine and others that scream that I must fight, flee, or exorcise the pain by ending it all.

Mark says this is just an excuse because I don't love him. My primary care doctor concurs in her way. She says I "somaticize too much" and suggests that my poetic imagination is to blame for my pain.

AD 1486

German Dominicans, perhaps inspired by fellow friar St. Thomas, publish *Malleus Maleficarum*, a veritable how-to manual for identifying and convicting witches. It takes one line from the Bible as its mandate: "Thou shalt not suffer a witch to live."

Malleus Maleficarum, or *The Hammer of Witches* as it is known in English, sets the stage for the witch hunts that sweep Europe for the next few centuries. Its four hundred–some pages boil down to this decree: if you are ill and you are a woman, and a doctor cannot determine what is wrong with you, then you are probably a witch, and the Devil is to blame.

What follows, if not exorcism and flogging, is execution: burning at the stake, along with less storied forms of murder, including hanging at the gallows, strangling, and beheading.

By the eighteenth century, as many as one hundred thousand Europeans are killed for conspiring with the Devil, acting hysterical, healing others, and various other documented manifestations of witchcraft. At least 80 percent of the murdered are women. Most are healers, peasants, single, or elderly, and many were ill, in mourning, or the victims of chronic abuse.

AD 2003

New patient: Jennifer Soriano. Twenty-seven-year-old Filipina. Nonprofit employee. Second time seeking therapy. Cites history of depression. Previously unable to find suitable therapist. Last therapist (Susan X) offended patient with theories of Asian heritage and taboos, suggestion of sexual abuse with no basis.

Jennifer complains of persistent pain in sacrum, arms, neck. Excerpt from journal, which she brought to first session:

> *I feel crazy. The pain is becoming me. It is taking over me like smoke subsumes a room on fire. My head shakes left and right involuntarily, as if I'm saying no to something inconceivable. My ankle twitches, my pelvic bone cracks, I swallow a rising howl. I shake my head and fix my face. I prepare myself for another day of pretending to be good.*

Pain may be classically psychosomatic, emotionally based, physically experienced, real. Says she hears voices. Voices say variations of no, express anger, ask questions. None urge her to cause harm.

Therapies to explore: motherwound, teddy bear technique.

AD 1663

On the Philippine island of Panay, a spiritual healer called a babaylan leads a revolt against Spanish colonizers and their weapon of pacification, Catholicism. Colonizers brand the babaylan, named Tapar, a "sorcerer" and "priest of the demon" and describe them as a man who "goes about in the garb of a woman." Tapar is accompanied by María Santísima, who perhaps is also a babaylan herself. Spaniards label her a prostitute and believe she is mocking the Blessed Virgin by adopting her name.

Tapar and María Santísima burn the village church, then flee to the mountains. Colonizers eventually catch them and have their revenge: they fasten the witches to bamboo stakes at the mouths of the rivers Halawod and Laglag, where crocodiles are said to have finished them off.

AD 2008

J. Soriano, thirty-two years old, fifteen-plus years of chronic pain. Accumulated patient inventory to aid with physiological malaise:

- Wartenberg Wheel and Tomahawk Hammer
- Transcutaneous electrical nerve stimulation (TENS) machine with wires and electrodes to interrupt habitual pain transmission with battery-powered pulses
- Thera Cane massager for muscle spasm tension
- Plastic traction machine with inflatable polyurethane pillows to mold missing curves into spine
- Rubber-studded balls to prod feeling into Bermuda Triangles around sacrum and neck, where nerves have adhered to masses of muscle, tendon, and bone
- Sacroiliac belt to stabilize weak and misaligned pelvis
- Double-domed polyurethane trigger-point prop to relieve pressure around cranial nerves
- Ace bandages for tendonitis in ankles and wrists
- Paraffin bath machine to coat hands and forearms in hot wax for pain relief and treatment of muscle spasms
- Cortisone shots for neck and sacrum
- Trazodone and Zoloft for secondary effects of chronic pain relief
- THC tincture and cyclobenzaprine and tramadol—the good stuff, what really works, what calms the nerves and temporarily finishes off the pain

AD 1878

Neurologist Jean-Martin Charcot ushers in a new era by coining the term *traumatic hysteria*. Case histories detail the connection between "psychical shocks" and physiological malaise. Through these histories, Jean-Martin becomes the first to conclude that women with unexplained illness are not at all possessed by the Devil, nor are they simply overemotional; instead, they suffer from both the mental and physical aftereffects of a traumatic event.

This is groundbreaking. But Jean-Martin goes further; he declares that hysteria can also affect men. Abnormalities of the central nervous system can lead to convulsions, trembling, and fainting. He becomes one of the first to show that psychological trauma can have a physiological basis in the nervous system, regardless of gender. Most of his colleagues are not ready to hear this. Perhaps if Jean-Martin were a woman, he would have been burned at the stake.

AD 2010

Patient, Jennifer S. Thirty-four-year-old female. Self-referred for an MRI and a neurokinetic evaluation. Conducted needle electromyography. Electrode inserted in all target areas. Motor neuron activity normal. MRI shows one abnormality: an arachnoid cyst to the left of the sacrum.

This arachnoid cyst may be a source of her pain. But it is simply a part of her anatomy now.

AD 1880

Viennese physician Josef Breuer begins treating patients with a technique influenced by Jean-Martin Charcot. Like Jean-Martin, he believes hysteria results from psychic trauma. Unlike Jean-Martin, he believes the cure lies in talking it out.

Josef begins to talk it out with a woman named Anna O. Anna O. suffers from paralysis, disturbances in vision, speech impediments, and a number of other symptoms, including hallucinations. Josef encourages Anna to express her feelings and to explain what hurts. Over time, some of her symptoms begin to subside. Anna O. calls the method the "talking cure," which she also refers to as "chimney sweeping."

AD 2011

I find a new therapist to talk it out with, except instead of talking, she asks me to write it out. She gives me an exercise called La Niña Interior. I write questions to my inner child with my right hand and respond as the inner child with my left.

How are you today, niña interior? Who is with you and what do you need? All the answers I scrawl with my nondominant hand are meant to surface the buried memories of my inner child. All the answers I scrawl amount to my inner child feeling alone and afraid.

AD 1895

In 1895, a young student of Josef Breuer's named Sigmund Freud convinces Breuer to write up the case of Anna O. Josef and Sigmund publish a paper called "Studies on Hysteria." Because such celebrated doctors have written her story, Anna O. also becomes celebrated—as the first successful psychotherapeutic patient.

But it turns out that Anna O. was not cured by talking alone. When Josef ends her therapy, Anna O. survives, evolves, and eventually thrives thanks to the support of her cousin, Anna Ettlinger. Anna E. is by no means a scientist, doctor, psychotherapist, or even a priestess or healer of the traditional kind. Anna E. is a writer.

AD 2012

I write in my journal about an old dream: I am in the Philippines, watching my great-grandfather pull a net from the tilapia pond in his backyard. He catches a fish and tosses it to his wife, who catches the silver creature in her mouth and swallows it live, whole. A mouse scampers onto her bare foot. It stops to nibble her toenail before hurling itself into the pond.

In the distance, there is rumbling and several sharp pops. The two turn their heads, and when they turn back, they have become my grandfather and grandmother, the tilapia pond is now the main market of Manila, and they are running from an advancing troop of Japanese and American soldiers. The market floors bleed. A woman's hand with delicately filed nails lies by itself on the ground.

I have had this dream since I was a child. More than three decades later, I realize something new: every time I wake from this dream, my hands are throbbing. I am clenching and twisting and wringing them, as if using every ounce of strength to live, to keep my body parts from flying away.

AD 1936

Bertha Pappenheim dies. Who? Bertha Pappenheim, the real woman behind the invented patient named Anna O.

Before she died, Bertha Pappenheim—whom Josef and Sigmund named Anna O. to protect her privacy and also her shameful diagnosis of traumatic hysteria—transformed herself despite sickness into a remarkable women's activist and author.

Bertha was ill for seven years after Josef Breuer ended her talking-cure therapy. But thanks to the support of her writer cousin, Bertha published her first collection of short stories, and gradually after this accomplishment, she recovered. During the last four decades of her life, Bertha published a prolific amount of work as part of her activism in the women's movement, including another collection of short stories, a German translation of Mary Wollstonecraft's *A Vindication of the Rights of Woman*, and an original play called *Women's Rights*, in which she criticizes gender-based economic and sexual exploitation.

Apparently, for Bertha P., the writing cure proved even more therapeutic than the talking cure.

AD 2013

I scribble notes on what I've learned about in vitro fertilization and wait patiently for the doctor to say something. She moves her reading glasses to the precipice of her nose and examines a gray film at arm's length.

Silence. She shifts in her chair, flicks her hands so the film goes *thwack*, and moves it up toward a sickly florescent light.

"Your uterus is tipped," she says finally. "And see this?" She points her pen at a cloudy, curved line. "That's your fallopian tube."

I squint at the ghostly image. If she weren't narrating this show-and-tell, I would think I was looking at paranormal activity caught on film.

"But where is your ovary?" the doctor continues. "It must be hiding behind your uterus. And see the other one? The other tube is like a Slinky."

My husband looks at me like he's afraid I might faint. Instead, I bow my head to suppress a chuckle. Leave it to the fertility doctor to discover what I had long suspected: something is amiss in my womb.

I smile because it's amusing to learn there is so much interesting activity down there, in an area that always felt comatose, near dead. I have never fully identified as a woman with a womb. And so I am entertained by the thought of these reproductive organs acting so feisty and mischievous—the fallopian tubes contorting like acrobats, the ovaries and uterus playing hide-and-go-seek.

AD 1980

The American Psychological Association eliminates hysteria as a pathology, ending, one would think, the historical obsession with a misbehaved womb. Hysteria becomes divided into various descriptive diagnoses, including dissociative identity disorder, somatization disorder, acute stress disorder, fibromyalgia, and PTSD.

On the one hand, this is a victory for people who have advocated for serious treatment of debilitating chronic fatigue and pain. On the other hand, Western medicine in all its glory remains stumped by systemic conditions like these, whose adequate treatment requires an integrated and nonbinary view of mind and body, which is to say, an integrated and nonbinary view of the world.

AD 2014

"I think I have fibromyalgia," I say to a doctor friend of mine.

"Oh no, really?"

"Well, my doctors keep asking me if I have it, so maybe I do."

"Honestly, fibro patients are the worst." He rolls his eyes. "I just never know what to do with them."

"I can't believe you just said that to me."

"It's true." He sighs, as if he is the one in pain. "I don't know what to tell you. Maybe you should meditate."

I am not this body. I am universal consciousness, no color, no gender, no shape.

I am this body. I am not universal consciousness, no color, no gender, no shape.

I am this shape. Separate color. Many gendered. Body conscious. Not universal, no.

AD 2011

TIME magazine publishes a cover story on pain. The news is that chronic pain is now treated as its own disease, not just a sign of psychological deficiencies or a symptom of other physical disorders like misplaced wombs. The possible causes of chronic pain disorder are myriad.

Ten years earlier, pain was included in medical protocols as the fifth vital sign, alongside blood pressure, respiratory rate, heart rate, and temperature. Around the same time, the Girl Who Cried Pain study found that women are more likely to be prescribed sedatives or antidepressants than men, who are more likely to be given stronger and more effective pain medication.

One in five people across the globe suffer from chronic pain, most of them women. Growing evidence shows that women of color are the most likely of all populations to be undertreated for pain.

AD 2014

Audre Lorde on pain:

> *Pain is important: how we evade it, how we succumb to it, how we deal with it, how we transcend it.*

Aurora Levins Morales on pain:

> *The only way to bear the overwhelming pain of oppression is by telling, in all its detail, in the presence of witnesses and in a context of resistance, how unbearable it is.*

Eula Biss on pain:

> *If no pain is possible, then, another question—is no pain desirable? Does the absence of pain equal the absence of everything?*

Audre Lorde again on pain:

> *I had known the pain, and survived it. It only remained for me to give it voice, to share it for use, that the pain not be wasted.*

AD 2016

What is the use of pain when, like history, it repeats itself?

I am awake at 2:30 a.m. wondering.

It is another night where pain's insistent whispers are louder than the beckoning of sleep.

Pain begs for a solution, but solutions are not always the answer. I think back to when I was finally prescribed opiates. The pills brought me into a state of analgesic bliss. But when I crashed off the high, I felt the absence of everything, and that void was far worse than the pain.

Now, instead of quick solutions, I seek simple acts of transcendence. I get out of bed when I can, which is a chronic act of faith. I live from hour to hour, which is the practice of creating history. I write in cracks of free time, which is the ritual of growing our mothers' gardens. I ask questions, which is the pilgrimage toward knowing I am not alone.

> *What if pain were a measure not of illness but of human need? What if we embraced rather than disposed the broken and the lost, the mysterious and misbehaving and misplaced?*

When pain keeps me up at the witching hour, I wonder about the healers and the hysterics. What would my story be if I had lived in 500 BC? In 1800 or even 1960 AD? What would their stories become if they existed in their bodies in our time? How many witches would have lived to write, how many hysterics would have healed to tell.

I
Neurogenesis

- Neuron creation that continues through adulthood but is especially dynamic in utero through teenage years.
- The origins of heritage, in which I seek the source of my chronic pain and what shaped my developing nervous system.

I can believe almost anything—

that we began
as thoughts an ocean away   *carried as seeds or smog or trash*
across the water

—MICHELLE PEÑALOZA

Nervous

My father raises his arms and shakes them once, as if flicking water from his hands. He looks like a magician, ready to perform an amazing trick before my very eyes. He occupies the center of the operating room, dressed in green scrubs, a bouffant cap, and "flesh"-colored surgical gloves. Meanwhile, I am standing in a corner, fidgeting with my pink-and-white candy striper apron, rubbing the rough seersucker between thumb and forefinger, again, and again, and again. To say that I am nervous is an understatement. In my fourteen-year-old mind, this moment is as critical for me as for the patient knocked out on the operating table.

My father has finally deemed me mature enough to witness the work that feeds our family of five. Before this, he had mostly left me to my own devices. He and my mother are more focused on work and each other than on connecting with their kids, so I have learned to mold myself into the strong-and-silent type in order to receive their approval, if not their attention and affection.

Here in the operating room, my father's domain, I worry I won't have enough mettle to keep my act up through the sight of cut flesh. But I remind myself that I grew up perusing photo albums full of exposed brains and playing with model skeletons and skulls. And so I stand taller and tell myself I can handle it. But can I?

My father uses his upper arm to nudge his horn-rimmed glasses back up the bridge of his nose. Then the surgery begins. With his left hand, he picks up his first tool, a blue plastic razor, which looks to me like the kind you can buy at the 7-Eleven down the street. I step forward from the corner to get a better view but stop when the nurse looks back at me and glares. I retreat a little, mortified, and hope that my father hasn't seen my misstep.

The patient's hair falls to the floor. As it does, I picture my father applying the razor to shave his own head. Maybe this is why it looks the way it does: with only a few strands of hair left to comb over a bald dome. I suppress a tiny chuckle, then silently scold myself: *Stop being disrespectful; focus on the procedure.*

My father shaves the patient's hair like he's mowing a tiny lawn, and when the last patch falls to the ground, he trades the razor for a Sharpie, and draws two dotted lines across the patient's scalp. The place where the lines intersect is an *X* that marks the spot of where to cut. Now the nurse gives my father a plastic-handled sponge coated in brown liquid. Some of the liquid drips to the floor, and I wonder who will have to clean it up. He takes the sponge to the patient's now smooth dome and mops till it shines like a caramel apple.

I shift my weight from left foot to right and back again. Even though I know this is a routine surgery, I can't shake the feeling that something will go wrong. After all, I'm a sickly and accident-prone kid—what if my bad luck causes the heart monitor to flatline, as seen on TV? To calm myself, I focus on my father's confident hands, matched only by my confidence that something will go awry. For reasons I don't yet understand, I am always prepared for disaster.

My father crosses himself and says a short prayer. He balances a scalpel between thumb and forefinger and brings its glinting tip to the patient's scalp, then slides the blade across the skin, leaving a line of blood in its wake. He pins back the skin flap to expose the patient's skull. Finished with the incision, he exchanges the delicate scalpel for a drill as big as his forearm.

I am still standing in the corner, but in my mind's eye, I see the patient from above, as if I'm floating just below the drop ceiling. I wonder if the patient is really asleep under the green drapes that cover his face and body, or if his eyes are open and he can secretly feel his head being sliced.

Suddenly, there's a pang in my stomach, and I double over ever so slightly. What is this pain? Is it the fear of disaster, or nausea from seeing blood, or just one of those strange jolts that sometimes shoot through my body?

The surgical nurse looks at me again, this time with concern rather than reproach. "You gonna be okay?" she asks.

I force a smile and nod, disappointed in myself for betraying weakness. If the patient on the table can be still, then I can too, I think, even though anesthesia makes this a ridiculous comparison.

"You should sit down in case you get faint." She waves at a stack of foam boxes piled behind me in the corner.

I shake my head, and she looks at me hard. I think she's waiting for me to speak, but I'm afraid if I open my mouth, my voice will shake and betray how nervous I actually am. What if she kicks me out? What if I embarrass my father, and he regrets inviting me here? What if he sees my true immaturity and further dismisses my worth?

None of these options is acceptable to me, and so I stay silent. Another pang shoots through my abdomen, but this time I stay upright and keep my poker face strong.

In my father's sure and ambidextrous hands, the drill whirrs and buzzes, the bit spinning like a tornado as it moves toward the patient's exposed skull. The drill makes contact, and bone chips fly. The chips look like soap flakes but hit like fingernails. Some stick to my apron, and I quickly brush them away.

There is now a quarter-size hole in the patient's skull, and the nurse cleans it by sucking away the remaining bone chips with a vacuum tube. My father exchanges the drill for a scalpel once again. He pierces through a white membrane called the "dura mater," the tough mother, and I am equal parts amazed and disgusted to see blood pour from the incision like watered lava from a volcano.

The nurse irrigates the opening and hands him another tube, which my father positions inside the hole to remove a blood clot, the objective of this procedure. The vacuum tube goes *whoosh* and sucks the clot from the space between brain and skull.

Wouldn't it be nice, I think, *if there were a vacuum tube to suck away my pain?*

It's become more difficult to focus on the operation; electric sensations have turned my arms into lightning rods and eclipsed the now murmuring stomach pangs. I bite my nails against the pain but puff out my chest to make it look like I'm standing strong and at attention. As my father exposes a brain and reveals his expertise on the nervous system, I am learning to conceal what is going on inside my body. This is the beginning of my own expertise as a performer, skilled in prestidigitation to hide pain and in the magic of making my vulnerabilities disappear.

My father stitches the patient's skin back together, then adds staples to seal skin over the hole. The skull bone will grow back over the course of a few months. In my head, I wish the patient well and hope for a full recovery.

My father rubs his gloved hands together to signal that he's all done. Then, to my surprise, he whistles: four tones, two high and two low, like a doubled doorbell ding-dong. This is the signature tune he performs when he comes home from the hospital at a decent hour, to signal to our family that all is well. After the whistle, he waves at me as he exits the room to scrub down.

I release a breath I hadn't realized I'd been holding. The patient made it through the surgery, and I did as well. My father's good-natured behavior must be a sign of approval; I was a smart observer and a stoic witness, and I had done him proud. I smile and mentally pat myself on the back even as the electric currents and stomach pangs endure.

The nurse escorts me to the bathroom, where I wash up and pick a stray bone chip from my apron, then return to my candy striper job down the hall. I think about my father's steady hands as I file manila folders and watch my own hands shake.

~

A week later, my father and I go out to dinner, just the two of us—a rare and awkward occasion. He chooses a place called Toby's, a white tablecloth joint adorned with beer steins shaped like trolls.

We don't talk about the surgery then, or ever. Witnessing it had a sort of *ta-da* quality to it, as in, *Ta-da, there it is!* and then the star of the show disappears behind a curtain. I was left to process the significance of the performance on my own.

At dinner, I twist a napkin over and over again in my lap, looking intently at my food, working harder at finding words than at chewing the tough meat off my plate. My father remains focused on eating. He cuts his steak with surgical precision. He loosens the knot of his tie, perhaps to allow more neck room to accommodate generous spoonfuls of meat and yellow rice. He seems unperturbed by any need to speak. I should be used to this silence by now, but sitting with it, literally face-to-face, has made the stomach pangs return. I continue to chew my meat, even though I've lost my appetite.

~

Growing up, I rarely interacted with my father because he was always working. I knew him as a disembodied voice over the phone when he would call in the evenings or on weekends to check in from the hospital. I knew him as the indentation in my parents' bed where he'd thrown off the covers to get to the operating room before dawn.

I mostly remember relating to my father through surgical-like procedures. When I was three, he pierced my ears using a needle and ice. I laid down on the coffee table while my brothers and my mother surrounded me and held my arms to make sure I didn't flail. At that time, I also saw myself from high above. My eyes floated by the ceiling while my body remained on the table as my father incised one hole into each earlobe. I was frightened, but I don't remember feeling any pain. Afterward, I ran upstairs to look at myself in the bathroom mirror. Staying silent while scared had its payoffs. I was pleased with the pinhead-size turquoise balls that now peeked below my bowl-cut hair.

When I was eight, my father removed a baby tooth from my mouth by burning through my gums. A strip of gum tissue had grown over the baby tooth like a pink tether, because I was missing a permanent tooth underneath to push it out. My father applied Anbesol with a Q-tip to anesthetize my gums, then used a cauterizing tool to burn away the tether. He touched the metal loop to the strip of gum, and the heat radiated to my tongue and inner cheek. I prepared myself for agony but felt no pain. The smell, though, was awful. As smoke curled out of my mouth, my nose filled with the scent of burning flesh. My father continued applying pressure with the cautery tool until my tooth clattered into the sink. I ran my tongue over the hole where the loose tooth used to be and felt satisfied.

Afterward, I told this story to friends at school. I liked hearing how grossed out they were and enjoyed showing off that I had a father who could operate on me at home. If I couldn't feel comfortable around my dad because I barely knew who he was, I could at least feel proud of the cool things he knew how to do.

~

And then as I grew older, things started to happen that no surgical procedure could fix. My ankles began to tremble when I walked. I ensconced them in thick wraps of medical tape, and when unwrapped, they would shake for hours. By the time I was sixteen years old, I had to quit the sports I loved. My right kneecap would dislocate in the middle of soccer practice. A vise grip of pain in my neck and arms prevented me from holding a tennis racket for the course of a match. If I moved my head too quickly to get my eyes on the ball, my neck would stick in a twisted position until I snapped it back.

My high school tennis coach urged me to get my neck checked out. Judging from the look on her face when it happened, I think she was grossed out by seeing my head torqued like that. And so I got an X-ray of my cervical spine, which showed a reverse curve and compressed discs in my neck. The doctor recommended cortisone shots to reduce the chronic swelling that likely contributed to the problem.

One Saturday afternoon, when my father had a rare break at home, I gathered up the courage to show him the X-ray. I sidestepped over to where he lay reclined in his Barcalounger, watching *M*A*S*H.*

"Hey, Dad," I said in a small voice. "Will you look at these?"

I held the X-ray in front of him, and without touching it, he gave a cursory glance. "It doesn't look bad," he declared. "You don't need surgery." His eyes went back to the projection TV, where Dr. Hawkeye was speaking to a skeleton.

My face fell, but I steeled myself to engage him again. In an even tinier voice, I asked, "The doctor said I could get cortisone shots. Do you think I should?"

"The shots will help relieve the pain," he replied briskly, still watching Hawkeye on the screen. "But they're temporary and won't solve anything. It's probably better to just tough it out."

This felt like a slow and suffocating squeeze to the heart. Not even this visual evidence of a problem in my neck was enough to warrant sustained attention from my dad. I felt dismissed, unworthy of care. I wanted to scream, *I don't want to tough it out!* I longed for softness to buffer the rebar in my neck and the nail files in my back; I wanted to be hugged and cooed at and coddled—yes, even at sixteen. Instead, my father's clinical distance and minimization, which was mirrored by my mother as well, sparked emotional pain that felt worse than the physical aches. This layering of both physical and emotional hurt was too much for my teenage body to bear. And so instead of screaming, I locked down my emotions and tried to adopt an equally dismissive approach to my pain.

~

But inside, my body was ringing an alarm. I clenched my muscles to keep my joints from giving way. I braced myself against unpredictable currents of electricity in my arms and back and numbed out against the relentless ache that gripped my shoulders and sacrum. I contracted myself until, like a pill bug, I felt I could either play dead or roll away.

This chronic clenching helped me get through the day, but at night, when it was just me unfurled in bed breathing, the pain crept out of hiding. It seemed to seep like smoke into all voids in my body, and it forced me to face my own tenderness.

In response, I sent my mind to battle. Whenever my father faced a challenge, he would say in his Tagalog-inflected English, "I'll gonna do it," and inevitably he did it. Meanwhile, I'd think. *I'm gonna do it*, and then I'd climb into bed overcome by fatigue. "Tough it out," I'd whisper to myself while horizontal. "It's not a big deal. You just have to be stronger."

My body won out over my will every time. The sensation of rebar in my neck seemed to jab further into my cranium, and my sacrum pain would grow into a five-alarm fire. The pain would keep me up for countless nights despite my resolve to overcome it. Why couldn't I just think these ailments away?

~

When we think about the nervous system, the first thing that comes to mind is probably the brain. The second thing might be all the conscious things the brain can control, like eating a sandwich or performing brain surgery. Most of us were trained to view the nervous system like a top-down, brute-force machine fueled by conscious choice. My father and I were no exception.

My father learned these dominant views in medical school in the Philippines and also when he trained as a resident in the US. He ascribed to a headstrong view of life. In addition to saying, "I'm gonna do it," he quipped phrases like "Where there's a will there's a way," "Mind over matter," and "Punch first before the other guy punches you." That day in the hospital, I saw how this view applied to the way he operated.

When I witnessed my father perform surgery, I was extremely self-conscious, too worried about doing the wrong thing to see the importance of the moment. Looking back, I now appreciate it as the only time I got to see my father's lifesaving skills with my own eyes. It was also the first and only time I was afforded a window into my father's approach to the nervous system. He was a confident biomechanic who used his will to manipulate tools to solve concrete problems.

In college I studied the history of science, and it was then that I began to see my father's approach to the nervous system in a larger context. I learned that in the nineteenth century, Western views of the nervous system compared it to railroads and the telegraph, two tools necessary to achieve manifest destiny and territorial expansion overseas. In the second half of the twentieth century, the advent of computing led to comparing the nervous system to a mainframe, with massive data processing capabilities, inputs and outputs, and the capacity to run multiple programs controlled by the main "brain."

It strikes me that railroads, telegraphs, and old mainframe computers are all systems of rigid infrastructure. They are static and resist—rather than work with—the environment, so that the hardware can carry changing loads, communication, and software with minimal breakdowns. In general, these systems are either up and running, or shut down; they are working, or they are not—there is rarely an adaptive in-between.

That day in the operating room, I saw my father use a razor, a scalpel, a drill, and a stapler to bring a system under threat of shutdown back under control. He was precise and efficient, just as a rail-track mechanic must be with their wrench, tongs, shanks, spikes, and bolts in order to avert disaster. This approach saves lives, but it simply didn't apply when my own nervous system began to go awry.

~

While training in the Bronx, my father saw an extraordinary amount of acute physical trauma: pierced skulls from blunt force wounds, backs impaled with knives and screwdrivers, heads hemorrhaging from beatings, bullet shards embedded in brains. For two years, my father operated regularly on injuries like these and was often called back for surgery even after his long shifts were done.

This intensive training prepared him for what was to come. When he and my mother moved from New York to Chicago, he became the main trauma neurosurgeon in southwest Chicagoland. At the peak of his career in the 1980s, he worked at five hospitals, including Christ Hospital, which was one of the only trauma centers serving the entire South Side. At Christ he operated on what he called "a lot of gang-related gunshot wounds to the head."

One evening close to midnight, he was leaving the hospital after a full day of procedures when paramedics rushed in with a bleeding patient. It was yet another victim of a gunshot wound. This bullet had lacerated the patient's transverse sinus, where fluid drains in the cranium, causing blood to spurt like a fountain from the back of the victim's head. My father remembers using his fingers to plug up the blood while he rushed that patient into the operating room.

This was the nervous system my father was trained to heal. A nervous system in acute crisis from physical trauma and also, as I now see it, from racialized economic divides. His way of seeing the nervous system allowed him to save lives under pressure; it also obscured my own growing pain from registering as anything that required attention.

~

When I was twenty-three years old and newly moved to San Francisco, my parents came to visit me in my adopted home. I took them to Golden Gate Park and showed them my running trail. Seven years after I quit team sports, I had become addicted to running. Three times a week, I ran a six-mile loop on dirt trails to the windswept expanse of Ocean Beach and back again to my Sunset neighborhood apartment. Along the way, my left ankle would twist when I stepped on a tree root, my right knee would throb, and by the end of the run, I felt like daggers were jammed into my hip sockets. But the endorphin rush I felt from the run outweighed the consequent pain. After all, I was always in pain. In my body-mind, gaining an hour of natural pain relief from a runner's high justified the beating I was giving to my tender joints.

That day, after a few hours of strolling through the park, a blast of fog overtook my brain. I could no longer keep my eyes open and told my parents I had to go lie down. It was part and parcel of my chronic pain to overtake me at odd times. Whereas I could blast through six-mile runs, pop seven Aleves, and keep going, it was common for my system to shut down after pouring cereal or going for a walk in the park.

To her credit, my mother told my father, "Go help your daughter," and this time, he did.

I lay in bed feeling like a bag of pus. He conducted an examination—the only one he ever gave me—by palpating my back, arms, and legs. My whole body was inflamed. My father asked me to perform a number of repetitive motions, including balling my hands into fists and turning my head from side to side. I weakly complied before closing my eyes. I just wanted to go to sleep.

His diagnosis:

"The pain is probably from what we saw in that X-ray you got in high school. The compressed discs in your neck may be putting pressure on the nerves. You also may have a mild form of dystonia."

I was surprised he remembered that X-ray. Apparently he had been paying attention. I was in awe of this, and the fact that my father was now sitting down long enough to take my condition seriously. I wanted the moment to go on and on.

"What's dystonia?" I asked.

"Muscle spasms," he replied, standing up from his chair. "You have dispersed neuromuscular spasms throughout your back and extremities, especially your neck and arms. They probably get worse through small repetitive motions and put stress on your joints."

I suddenly felt affirmed and seen. The diagnosis matched what often happened in my body, especially after typing and runs. It had taken twenty-three years of my life to arrive at this point. My heart began to glow.

"Muscle spasms," I repeated, still lying down. I didn't want to stir; any sudden movements might reveal this moment to be a mirage. "That makes sense, I definitely feel like my muscles are constantly in spasm. What do you think is causing it?"

"No idea," he said. Then he looked down at me and smiled. "Don't worry, it won't shorten your life. It's not a gunshot wound to the head."

My heart shut both its doors. With one phrase he had gone back to minimizing my pain. I went from feeling seen to being invisible again. But like a reflex, I stayed silent. I knew my father meant to make me feel better in the only way he knew how, and so for his sake, I swallowed the hurt again and smiled back.

~

I had spent much of my teenage and young adult life hoping my father could get rid of my pain like he got rid of blood clots and bullet shards in the brain. I also spent much of my life trying to hide the pain while also wishing it could be seen, especially by my parents. It's only now, looking back, that I am acutely aware of how much my own dismissal of my pain contributed to years of suffering that might have been mitigated, if not prevented. Truth be told, I minimized my pain as much as my father did. I tried to turn away and ignore it because, after all, it wasn't a gunshot wound to the head.

This is why I gave up on telling doctors anything about my pain. This is why I continued to run six miles a day and swallow Aleve pills like vitamins. This is why I continued to push myself to achieve to prove my own worth, and why most of my friends never knew I was in pain.

But by my late twenties, the pain had grown to a point where I could no longer hide it, and for this I am now thankful. While I remained silent, the pain grew to a deafening volume inside me, until one day I tried to end it all. A few years after I came back from that precipice, I became so ill I had to take six months off work and put myself on self-administered disability leave. It was during that time that I finally owned up to the fact that no one else, not even my father, was going to save me.

Over the next fifteen years, between the ages of twenty-eight and forty-three, I assembled a hard-fought collection of diagnoses: peripheral neuralgia, mild scoliosis, dystonic neuromuscular spasms, central sensitivity syndrome, general anxiety disorder, social anxiety disorder, and complex PTSD.

As they amassed, I found these diagnoses affirming on the one hand and confusing on the other. They were affirming because I finally had names to justify the pain. But names are not root causes. What was so wrong with me that I had all these disparate issues jumbled up inside my body? My brothers didn't seem to suffer from anything of the sort, my father seemed healthy as a coconut, and as my mom liked to remind me, she'd never had any of these issues at all. The diagnoses, at first, just gave me new labels for my own failure.

Then one day, while reading up on the various conditions, I realized that they all had one thing in common: they were all issues affecting the nervous system. Suddenly the question that occupied me transformed from "What's wrong with me and why can't I overcome it?" and also from "What's wrong with me and why can't my father solve it?" to "What happened to my nervous system and how can I understand it?"

My father couldn't see my conditions because they are invisible and systemic. He was right in that I did not have life-threatening physical trauma—I did not have a gunshot wound to my head. But he didn't know about the psychological trauma that was intimately connected to my conditions. He didn't know that psychological trauma lives in our nervous systems, which means that it not only lives in our bodies, but affects our body-minds as a whole. And he was wrong when he said my conditions wouldn't shorten my life. Without proper treatment, the systemic effects of complex PTSD, central sensitization, dystonia, and neuralgia can wear away at your health like water can erode earth over time.

Psychological trauma cannot be sliced, spliced, drilled, or vacuumed away. But it can be treated with longer-term care, with a softer and more integrated approach than surgery. This approach begins with a radically different perspective on the nervous system.

The founder of modern neuroscience, Santiago Ramón y Cajal, viewed our nervous systems in terms of the natural world. Inspired by his upbringing in the preindustrial Spanish countryside, he described nerves as a "forest of outstretched trees," "tangled jungles," and "delightful gardens." He called neurons "the mysterious butterflies of the soul." This perspective led Cajal to think of the nervous system as organic and flexible, as far back as 1894. Today, this concept is known as nervous system plasticity.

Cajal asserted that the nervous system could not be rigid like a railroad or telegraph system, and that instead it must be dynamic and adaptive to the environment. He knew that the nervous system's capacity for fluidity was crucial to our survival.

~

In our nervous systems, water molecules diffuse like tributaries flow toward rivers. Diffusion tensor imaging, or DTI, creates brightly colored maps of these water molecules' journeys through the brain. Tracing these tributaries provides an approximate map of the neuronal connections that play key roles in the integration or disassociation of different parts of the mind. This water-mapping has shown significant differences in the tract connections of people with PTSD, anxiety, and central sensitization. Researchers are still unable to interpret what these differences mean; they just know that the unique topographies—the twists and deltas, the dams and blocks, the diversity in the nervous connections—are all there. These differences are part of our fluidity.

One day while scrolling through Instagram, I saw a composite picture of aerial shots of all of Washington state's rivers and creeks. I was amazed. The image looked just like a DTI brain map of neural tracts. From then on, rivers and nerves became part of my frame of reference for the nervous system, and eventually also part of my perception of ecosystems and societies as a whole.

When I was a child, I would page through my father's surgical photo albums and think the exposed brains looked like smashed macaroni with red sauce. Now I think of these same sulci and gyri as rivers carving through the desert, seen from ten thousand feet in the air. If I were to assemble an album of how I now view the nervous system after decades of tracing, research, and somatic therapies, it would look like an album of waterways. It would have glossy photos of neurochemical waves, tributaries of sprawling neural circuits, nervous impulses that travel in multiple directions like the ebb and flow of tides. It would feature captions that say, "Our nervous systems are more than just our brains" and "A healthy nervous system is like a healthy water cycle" and "I flow, therefore I am."

This perception of our nervous system as a system of waterways is metaphorically accurate to how it actually functions. Like waterways within ecosystems, our nervous systems work in close relationship to other systems in our bodies, to other people's bodies, and to the environments that envelop us all. We are meant to flow between nervous system states of sympathetic agitation and parasympathetic rest. And the vast majority of what our nervous systems do operates beyond conscious will, like gravity pulls on rain.

Bessel van der Kolk, the trauma therapist who authored *The Body Keeps the Score*, writes that trauma causes us to experience the world with "a different nervous system." Inside my body, my different nervous system feels like a river seeking its natural shape. No matter how I might seek to control it, my body wants to bend and twist and turn like a stream, to resist the imposition of being tamed. I am learning to embrace my body's wild beauty not as an aberration or weakness but as an adaptive strength that helps make me who I am.

This different nervous system is to me a post-traumatic nervous system, one in which our nerves integrate mind with body, a system that is not broken but that instead exists in an adaptive state between wellness and dis-ease. I have learned that the post-traumatic nervous system is worthy of care.

Matthew Cobb writes about the importance of metaphors as frameworks to not only describe the nervous system but also to evolve our understanding of how it works. If our nerves are more like trees and watersheds than telegraphs and railroads, then our post-traumatic nervous systems may be more responsive to gentle shaping over time rather than the blunt-force engineering of men. What kinds of care might spring from this approach, what new solutions to traumatic pain could flow from a softer and more generational view?

When my father was seventy-nine and I was forty-one, he died of congestive heart failure, a chronic and progressive condition which ultimately drowned his lungs. Not long before he passed away, I told him I was writing a book about the nervous system. I didn't tell him that the book might cast him in a complicated light.

I showed him Norman Doidge's book about neuroplasticity, *The Brain That Changes Itself*, and told him I wanted to write about my chronic conditions and the way the nervous system can be fluid like water. He paged through the table of contents and raised his eyebrows and said, "Oh, Jenny, I think this will be a very great book you're going to write." His reaction was robust, genuine, and endearing in its enthusiasm.

This was his way of seeing me in the end. As someone with the potential to do great things, even though he was unable to see my vulnerabilities and needs. Though my suffering was largely invisible to my father, in the end we shared the perspective that our nervous systems matter. We just came to this common ground from very different points of view.

Later in life, I developed a friendship with my father, which grew from a love of music and a passion for having opinions about the world. He knew we disagreed on many things, but he would say, "As long as you can reason through your opinion, I will respect it." I'd like to believe he would apply the same approach to our different opinions on nerves. He might even have seen our varying viewpoints as complementary rather than in conflict, two channels through which to navigate the system that fed us, and that adapts to keep all of us alive.

Body Language

I could dance with you, if you want."

This is the invitation I'd been fishing for. I stand up so fast from my chair, it nearly falls over. I want nothing more than to dance with this dreamy-looking brown boy in a crisp guayabera the color of Carnation creamer. He seems so . . . cool, and intriguing, and he has these sweet thick lips that look like they'd be velvet to kiss.

I try to make my desire obvious by turning my body to face his and tracking him with my gaze as he walks across the hall. I don't like to think of myself as a coy little ingénue, sending implicit signals to try to get the man to make the first move. But here I am, doing exactly that. I probably look like a hungry coyote; I could be drooling, but I don't care. I am so magnetized by this person's presence, I need to be near him, even though before he offered to dance, we hadn't exchanged a single word.

We are now inches apart on the dance floor, and we serve each other these knowing smiles. He interlaces his fingers in mine and rests his other hand in the crook above my hip bone. His skin is warm against my chronically cold hands. Then, pulling ever so slightly with his palm at my waist, we begin.

This otherwise ordinary moment of salsa at a mutual friend's wedding reception feels deeply significant to me, like a courtship dance. Consciously or not, our bodies seem to be testing their compatibility. And my body likes what it senses in his: confidence, rhythm, grace, and a subtle brand of bossiness that makes me smile because I can tell it's a match for my own.

We haven't even introduced ourselves to each other, but I know who he is: Juan C., the younger brother of the groom's best friend. From our initial point of contact, our bodies seem to birth a new language, one we both implicitly understand. Over the years, our verbal communication will stumble to catch up.

Juan and I marry five years later, and soon after, we move to his home island of Puerto Rico. I am thrilled about the prospects. After twelve years in the wintered summers of San Francisco, I long for the freeze of chronic pain to melt in island sun. I hope Juan and I will dance salsa every day.

~

Like my ancestral homeland of the Philippines, Puerto Rico is part of the imperial body of the Greater United States. Shortly after Juan and I move, we settle in the colonial city of Viejo San Juan. In a place like this, colonization seems quaint, even beautiful. Candy-colored buildings flank cobblestoned streets. Hibiscus and jasmine perfume sandstone walls. Doorways are crowned with soles truncos—half suns of slatted windows that create intricate latticeworks of light.

Along with its beauty, Viejo San Juan is the embodiment of centuries of imperial defense. The city remains partly walled, flanked by the ruins of Spanish forts, spread beneath posts that fly both the Puerto Rican and American flags. It is here, in a cauldron of ghosts familiar with battle, that a massage therapist and air conditioning repairman named Freddy finds history lodged in my throat.

I am lying on a table in a century-old house, a white sheet pulled up to my chest. Mosquitos buzz languidly through the warm air.

Freddy applies a viscous, medicinal-smelling oil and begins to work it into the roped muscles in my neck. Suddenly, he stops.

"What does it feel like here?" he asks.

"Um, like I swallowed a baseball and it never came out." What it really feels like is continuous choking. Words, breath, pills, bits of masticated food often get trapped in that part of my throat, causing me to cough and gag just to get oxygen through and solids down the pipe. At night with nothing to distract me from the sensation, I often feel an agitated sense of despair. But there's no need to dredge up details with a massage therapist I've just met.

"It must be *very* uncomfortable," he says. "Your muscles are in chronic spasm here, and it feels like your neck is pushed off to the right."

"You got it," I affirm. I'm glad he can detect the structural reasons for my neck pain. This shows me Freddy knows what he's doing. I'm confident that I'm about to receive an excellent massage. But I do not anticipate what comes next.

"But my goodness, there's . . ." He stops talking for a moment, then continues, "There's something big that happened here, and your neck is holding on to that. It's something, well . . . frankly, it's something terrifying."

I swallow with effort as I feel a jolt move from my neck through my shoulder. I am both hopeful and fearful that Freddy will tell me what that something might be. But he continues to work my shoulders in silence, and I feel a sense of disappointment mixed with relief.

After the massage, we sit at the kitchen table, and as I write his check, I ask how he could detect this terrible sensation in my neck.

"I just feel things," he replies.

"What do you think it could be?" I feel ashamed for asking, and can't bring myself to look him in the eyes.

"I got the feeling that what was in your throat isn't from allergies or just knots or anything; it's trauma. From a long time ago, when you were a child. I could feel it in your shoulder, and then by the time I got to your throat, to your hyoid bone, it was terrifying."

I shiver. It must be at least ninety degrees, but Freddy's prescience is chilling.

"That's where the cry was. In your throat," he continues. "But your body will release it when it's ready. I'm getting goose bumps. The spirits are here, and they're taking care of you. It's already done. They're telling me to tell you: you understand the tools of the world but don't give it your soul."

I was both disturbed and intrigued by Freddy's interpretation of what my throat held. Months after this session, my body sent a louder message, an equally cryptic but more horrifying one. The experience sent me on a four-year journey of learning about the body language of implicit memory. It would lead me to a revelation about some of the earliest lived trauma that I carried in my nerves.

~

Just five days before my thirty-sixth birthday, I am at home relaxing in our apartment on Caleta de San Juan. I check my phone to confirm the appointment I've made with a naturopath the next day. This is my birthday present to myself and also a gift to Juan: a session that will prepare our bodies for trying to conceive a child.

I lie in bed with the doors closed and the AC unit on low, resting in the humidity and nursing my chronic fatigue. Juan is at a late-night work meeting, so I am alone with our loyal street dog, Kabu, who is curled up like a donut on her bed. I ponder the vaulted ceiling as lamplight stains the wall. Tired but comfortable, I listen to the sounds of the old city—my landlord cackling with a friend downstairs, stray cats howling in heat, cruise ships bellowing their horns.

Reaching over to pick up a book from the night table, my hands suddenly begin to jerk in midair, like two fish convulsing out of water. Then they become fists jabbing toward the ceiling—left-right-left. My legs begin to kick from the knees. I try to bring my limbs down, but they have a will of their own. I flail on my back like a baby in a temper tantrum, feeling the urge to plead for my life, then the need to flee. But I have no words to form, and my legs have no knowledge of walking. And so I stay lying on my back, kicking and swatting and thrashing my head, twisting my torso, shaking all over in silence.

This continues for what feels like hours yet must have only been a few minutes. When my body finally relaxes, the tears come like sheets of rain along with undulating waves of pain. The despair that overtakes me is crushing. I want the earth to open and swallow me. I feel through my entire being that someone had wanted me to die.

Kabu trots over and licks my hand with her sandpaper tongue. This wetness, this tenderness, brings a sliver of me back into the present. I manage to reach for my phone to call Juan, and in the smallest voice, summoned from across the decades, I ask him to come home.

"It was really, really scary," I whisper to Juan when he arrives.

Juan is having trouble understanding what happened, since I am unable to find the right words to describe the experience. All I can say is "I feel crazy," and "It was so scary," and "It was like it was really happening," and "What the fuck was it?"

Juan turns on some music, which he knows helps me calm down, and holds me in his arms while we sit on the bed. He rocks me gently, pulling me into a soft seated dance, and in his embrace, a current of peace washes over me until I fall asleep.

I sleep like the dead. When I wake up the next morning, Juan gives me a kiss on the forehead and opens the bedroom doors to let out Kabu.

"How are you feeling?" he asks.

"Pretty bad," I reply. I feel bruised and bedraggled, like I'd been thrown overboard and washed up onshore.

"So what happened last night?" Juan sits next to me on the bed. "You said you were kicking and crying?"

"I don't know," I say, hugging my knees to my chest. "I have no idea, but it felt just really violent and really, really old."

Juan looks at me, concerned. "Oh no. Violent like how? And what do you mean old?"

"I don't know," I say again. "Violent like I was flailing to get someone off me. Like they were trying to choke me. And I felt like, this is actually happening, except that I was really little, like a baby or a little kid. I couldn't even walk or get up to make it stop."

"Maybe you were dreaming?" Juan ventures.

"No, no, I was wide-awake," I say with as much vehemence as I can muster in my exhausted state.

I am emotionally wrecked and too tired to try to explain further. Once again, I have something to be ashamed of. What was this episode if not some strange possession or further evidence of my weak state of mind?

Juan wraps me in another tender hug and asks no more. I hold tight to absorb his peace through osmosis, then go to the balcony to marvel at the ordinary day.

The next weekend after the episode, I tell a friend what happened. We walk on the sun-soaked path of the Paseo de la Princesa, and she listens attentively while I struggle to describe the scene. In the end, she suggests it was just some weird physical tic.

"You have all these bodily shakes and things, right?" she says, making her case.

She's trying to ease my anxiety, I think. But her words do just the opposite. It reminds me of how my father tried to make me feel better by saying at least I didn't have a gunshot wound to the head.

"I do have aches and pains often, but trust me, this was really different."

"That's true, it sounds like it was really strange." She sucks her teeth and shrugs. "But you know, sometimes we read too much into things. Sometimes a China is just a China. Hopefully it will pass."

"China" is the Puerto Rican word for "orange," supposedly because Portuguese sailors had called them "naranjas de China" to distinguish Southern Chinese sweet oranges from more tart oranges from Spain. I think it's ironic my friend chooses this metaphor, since it seems to me that calling a China a China is an example of something not simple at all. Instead it's something that has layers of implicit meaning beneath the language, just waiting to be interpreted and told.

~

In my childhood home, I became proficient in body language because so much was left unsaid. If my father came home with his head held high, he was in a good mood. If he came home with his head bowed and brow furrowed, it was time for me or one of my brothers to give him a forehead massage and to brace for yelling when the hospital called him back to work. If my mother turned her body ever so slightly away from me, it meant she needed space and was busy and did not want to be bothered. If she closed her bedroom door in the afternoon, I understood she had a headache and I should leave her alone.

I've also learned to interpret body language because my own body is verbose. It yells inappropriately. Sometimes I ignore these loud pain signals till they pass. Other times I translate them into signs that I need to slow down. I have learned to speak back to these expressions of pain, to tell my body that I am not really in danger when my sacrum and neck begin to throb. I don't normally need to understand the cause of these flare-ups; I only need to manage and move through them, to help my body return to an inside-voice baseline.

This episode, however, is different. Failing to delve deeper into its meaning feels like a clear and present threat. The experience is so harrowing that I need to understand it. Could it be another piece of the puzzle of why I carry such pain and despair through an otherwise decent life?

By this point I had already been diagnosed with complex PTSD from childhood neglect. So one morning, while sitting at an outdoor café in Plaza de Armas, I look up "embodied flashbacks" to see if perhaps this was a post-traumatic episode of some kind.

I come across an article that uses the phrase *embodied flashback* as an example of implicit memory. Implicit memory is the type of memory that allows us to never forget how to ride a bike or how to dance salsa with a compatible partner. It is also the type of memory that conditions us to associate one thing with another through simultaneous pairing, like a bell and food in the case of Pavlov's dog. It is habitual, motor-based, nonconscious and nonverbal memory. And in the case of infants and small toddlers, it is the only type of memory that exists before we have words.

This article makes me wonder if perhaps I had experienced a reenactment of a preverbal memory from early childhood. As I read more, I learn about the hippocampus, a horseshoe-shaped structure located roughly in the middle of our heads. The hippocampus is a part of the brain that encodes and transmits long-term memories. It doesn't develop its full set of connections until about the age of seven. This is why adults can't (explicitly) remember anything about infancy. From birth to two or three, the hippocampus might already be stockpiling memories, but it doesn't yet have the pathways to move these long-term memories out of storage. Perhaps somewhere deep in my brain I had a long-term memory stored that would explain the phantom fight. But if it was there, I had no way of hauling it into consciousness.

In contrast, our amygdalae are fully wired at birth. The amygdala is an almond-shaped structure located next to the hippocampus, and it plays a key role in detecting and responding to threat. In adults, the hippocampus regulates the amygdala like a sluice can control the impacts of a flood. When we encounter a possible threat, the hippocampus gives us access to long-term memories that can modulate our amygdala's response.

For example, a dog barking might tell your amygdala: *Trigger a startle response, then mobilize the legs to run away!* But your hippocampus can help recall important memories: *You like dogs. Remember Wolfie, the sweet mutt you had as a kid?* This context allows you to orient to the present and realize that the bark belongs to a tiny chihuahua next door. Instead of running away, you might lean down to let the pup lick your hand as you walk by.

Psychologist Louis Cozolino explains that because the amygdala is fully "online" at birth, but the hippocampus is not, we are essentially wired to embody negative experiences in very early childhood. In other words, we are born wired to fight because we can't yet flee, probably because it helps us survive.

This research affirms what I suspected my body was telling me: that I experienced an implicit flashback of fighting to survive. But it also affirms that I might never know exactly why, since narrative recall from that early in life is neurologically impossible.

Implicit flashbacks have been called "retention without remembering." I had retained and suppressed this memory for more than three decades of my life. Even if I couldn't know the story behind the memory, I had to wonder: *Why was my body speaking up now*?

~

Studies show that people who have otherwise inaccessible memories of childhood adversity often have flashbacks to traumatic events when they are home alone.

Perhaps my body recalled this memory because I was not only home alone but home alone in colonial Puerto Rico, where I am either comfortable enough to have this memory thaw or where I feel cumulatively triggered by the battle history embodied by Viejo San Juan. Perhaps this history evoked a fight-or-flight response that is somehow remembered from my ancestors, and perhaps this response triggered the embodied memory of the earliest fight I had experienced myself. Or perhaps it's because Juan and I are preparing to try to have a child, and as I orient myself to this new endeavor, it is triggering early childhood memories of harm that prior to this I've been unable to face.

Maybe it's all of these things, or none.

One thing I know for sure: this flashback happened in part because I felt safe. The security of my relationship with Juan, the depth of our implicit communication, and the nurturing of his emotional intelligence created a safety net that caught me when I came out the other side. Without this safety net, I believe the memory would have stayed dormant in my nerves. Our bodies have a way of knowing what we can handle and locking away the rest.

~

Implicit flashbacks expressed through the body lend themselves more to interpretive theory than to factual translation. There is thankfully no Google Translate for this kind of body language, and there are no accreditation programs for getting certified in interpreting what the body says. The ways our bodies speak are as infinitely varied as each of our life experiences. This makes it all the more amazing when others can understand our body language as if it were a native tongue of their own.

Four years later and a hemisphere away, I meet someone who can read the patterns in my body, someone who unlocks a theory of what might have happened to me. After Juan's fellowship ended, we were essentially pushed out of Puerto Rico by its colonial economy, which provides very few living-wage jobs. Once we realized we couldn't stay on the island, we decided to move to Seattle to be near my brothers and their families. I will later believe that we moved so I could also meet this body language translator.

Seattle is a haven for water lovers like me and a breeding ground for body-based healers. Though I missed the warmth and culture of Puerto Rico, I embraced Seattle's water and its culture of alternative healing. It didn't take long for me to find a massage therapist whose monthly sessions were a great relief for my pain. Jean Clough was a wonder. After my eight insurance-covered sessions were over, I asked if I could pay out of pocket for more. In response, she declared, "You could. Honestly though, I can't really help you anymore. But I know someone who can."

I think some body workers would have taken my money and continued to give me massages, no questions asked. But while others were content to help me manage my pain, Jean actually believed I could get better.

I will forever be grateful to her for referring me to the practitioner who calls himself the Bone Whisperer. *How corny,* I thought when I looked up his website. But quickly, I became a believer.

~

The Bone Whisperer's office is on the first floor of a modest 1990s North Seattle townhouse. It is an unassuming place with worn carpeting and musty, lived-in smells. His old dog, Sky, blankets the floor like a small bear rug. Outside the sliding glass door, his wife often places an impeccably arranged vase of flowers bursting pink, yellow, and cobalt blue.

A mechanical table fills the center of the room, along with an industrial machine with gray paddles. An assortment of foam-wrapped wooden knobs takes over a whole shelf. Not one but a pair of skeletons huddle in the corner, and posters detailing the many layers of the human body, including the somatic and autonomic nervous systems, cover the walls.

In order to treat his patients, the Bone Whisperer requires a story about the physical trauma each person has endured. The story informs his working hypothesis of how force has entered and embedded itself in the body. He spends about two-thirds of each session assessing and developing a story-based hypothesis. Then, based on this hypothesis, he spends the remaining one-third of the session on manual adjustment of the soft bones, tendons, and fluid systems that he says unite our soma through tensegrity.

He begins his assessment by asking for the "full story" of what happened to me.

"I—I don't really know why I'm in so much pain," I say, feeling anxious. "No one's ever really asked me for the full story."

Unperturbed, he asks more specific questions: "Have you had any car accidents?"

"Two minor ones." I describe them to him while he scrawls notes on a form that features a black-and-white diagram of a naked man.

His next question catches me off guard. "Do you know if you were a C-section?"

"Yes, actually I was," I reply.

"Do you know if you were a forceps birth?"

"That I don't know."

He stops taking notes to look at me through narrowed eyes. "Yeah," he says, slowly nodding. "Your head is really stuck and moved off to one side."

He puts down the paper and pen and asks me to lie down on the examination table. He palpates my head, neck, and the parts of my shoulders and sacral area that have always bothered me most. These were the exact parts of my body that were ablaze during my implicit flashback.

"Interesting," he says. "Two of your cervical vertebrae are out of place, and your right floating ribs are jammed."

"Yes," I say eagerly. In under ten minutes he has confirmed what it had taken me years to find out. "I once had an X-ray that showed those vertebrae were compressing some of the discs in my neck."

"Mm-hmm," he replies, continuing to palpate. "That must be very uncomfortable in there. I think we can fix that."

I listen closely, intrigued.

"It's like your whole head is caught on your right clavicle. And your left shoulder is hiked up to compensate for that pattern."

I had long hungered for this kind of explanation. To hear the distortions in my body interpreted through a combination of anatomical and metaphorical descriptions was so affirming, I thought I might cry. In the past, chiropractors adjusted me, acupuncturists needled me, orthopedists and neurologists scanned me, and craniosacral therapists relieved my immediate pain. But no one had ever thoroughly translated the invisible and complex interior patterns in my body. The Bone Whisperer was speaking my language.

"I think you might have been a breach-type C-section, and they probably yanked you out pretty hard," he says, pronouncing his story-based theory. He presses lightly on my temples. "I can feel where they took forceps on either side and squeezed and tugged and maybe also yanked you out by the neck and arm."

I reach up to touch both sides of my head, and when I imagine the pulling he described, I feel a jolt of electricity run from my head down my spine.

"That kind of birth trauma causes all kinds of problems," he explains. "Your body just grows around the reaction. You clamp down to try to keep your head from being pulled off your neck, and that keeps your larynx and hyoid bones from getting crushed. But then that's where all the twists and turns come from, and it impacts your breathing and swallowing and thinking and everything."

A snippet of that Fugees song enters my head: "I felt he found my letters and read each one out loud." Other body workers in the past had dismissed these patterns as bad posture or "weird wiring" if they were able to identify them at all. One of my chiropractors even called me a "treatment junkie," presumably because he didn't really believe I needed structural help at all.

Meanwhile, here was the Bone Whisperer, not only believing that I was in pain, but fluently understanding the language communicated by the contortions in my body. The neuroscience of implicit memory had given me words to explain the embodied flashback, and now the Bone Whisperer was offering a narrative theory as to why it might have occurred.

Perhaps I had been born fighting for my life.

~

I was born on August 14, 1976, at Christ Hospital in Oak Lawn, Illinois, the same hospital where my father worked as a trauma neurosurgeon. I was a planned C-section because my middle brother had been delivered by emergency C-section, and in those days, there was no going back to vaginal birth when you'd already got cut open once.

My mother doesn't remember much about my birth because she was under general anesthesia, as was a common practice in the 1970s. Meanwhile, my father was not there to help remember. Given his work schedule, he was probably in the next ward slicing someone's head open while I was being cut out of my mother's womb.

I've attempted and failed to recover documents from Christ Hospital since they only retain birth records for twelve years. Christ was a trauma center that served much of Chicago's South Side, and so in his time there, my father operated on "lots of gang-related gunshot wounds to the head." Now I wonder about the climate of being born in this environment. I wonder if there was a charge of trauma that circulated all the way from the emergency room to the labor and delivery department. Or perhaps the labor and delivery department had its own traumas. My father had, after all, wanted to become an obstetrician, but chose neurosurgery after two near-fatal episodes while rotating as an intern in the obstetrics ward.

While anesthesia no doubt affected my mother's ability to recall details about my birth, I also wonder if perhaps she blocked these memories because it was a traumatic event for her, as medicalized birth can be for many. Medicalized birth has been shown to be disproportionately traumatic for women of color, especially Black and Native women, as well as LGBTQIA+ people.

Once in a while, a statement will escape from my mother's lips like an orphaned remembrance. About a year after the Bone Whisperer offered his theory about birth trauma, my mother came to visit. She sat at our kitchen table with her hands folded around a crumpled Kleenex. I was headed out the door to run an errand, and I told her I'd be right back.

"Yes," she said, as if continuing a previous conversation. "And they didn't let us see you."

The words seemed to come as a reflex, like a leg jerking in response to a hammer tap on the knee.

I turned away from the door and back toward her. "What was that?"

"They didn't let me and Dad see you. After you were born. I don't know why."

"Oh." I put my bag down and approached her. She was staring into the distance.

"The doctors?" I asked. "The doctors didn't let you see me after I was born?"

"No, they didn't." She shook her head.

"Why?" I asked. "For how long?"

"I don't know," she responded. She was still staring at a spot on the wall.

I got chills when she shared this. I wondered if she had felt alone during delivery, with my father absent and without the village that might have surrounded her in the Philippines. But instead of asking, I chose silence. A therapist once told me, *Be careful who you try to change because you never know what's holding them together*. I sensed that this was a highly sensitive topic for my mother. I did not want to pull the delicate string that seemed to bind her. It was not my place to make her fall apart. So I turned gently away to run my errand and to leave her body in peace.

~

Perhaps my first fight had been with the obstetrician who delivered me. Perhaps I had been so injured in the process that the doctors had to do some kind of quick repair. Today, physical injury impacts about two babies out of every one thousand born. Native American infants suffer from disproportionate birth injuries, and African American infants are victims of the highest rate of birthing deaths. Back in the 1970s, when forceps and vacuums were regularly used to tug out babies during delivery, I can only assume this rate, as well as the disproportional racial impacts, were much higher.

I asked the Bone Whisperer what he thought of my theory, that I might have been whisked away to triage to mitigate damage from delivery. He nodded and replied, "Oh yeah, that could have happened. It's possible. The way the patterns in your body are and how locked up it is around your lungs and trachea, you might have been in real danger."

Babies are malleable. Like our nervous systems are fluid. I may have been able to recover from this early trauma. But recovery from birth trauma requires being touched, soothed, and stewarded back to homeostasis by at least one nurturing caregiver.

They didn't let us see you.

My body recalls this absence and a fight to protect my life. These implicit memories were reinforced, through body language more than words, long after I came home.

~

I later emailed my mother, asking for her thoughts on why the doctors might not have let them see me right away. I relayed the Bone Whisperer's birth trauma theory and asked for her opinion. She responded almost immediately, "Regarding your birth, there was no trauma because I didn't go through any labor since you were delivered by Caesarean section."

I was familiar with this dance. The one where my mother would say one thing and then explain it away the next day. I wish I could have asked her this question in person rather than over email, so I could have at least tried to read her body language, which might have told me more than the words she actually said.

Omission

I toddle down the hallway to my parents' room, avoiding the steep flight of stairs along the way. Their door is closed and towering and has a knob far beyond my reach. I lift my arm and hit the dark wood with an open hand once, twice.

A moment later, the door cracks open and out comes a firm voice. "Dad is working and Mommy has a headache. You go play in your room now. You understand? Mommy has to sleep."

I look down at my little toes and extend my arms for a hug. The door closes with a *click*, and I am left standing alone. Though I have few words to speak, I understand. I shuffle back to my room, where silence, like an old friend, embraces me.

This is an adaptation of one of my mother's favorite stories. She recorded the anecdote in her journal. When I was a young adult, she photocopied the entry and mailed it to me with a note: "Look what I found. You were only two and you were already so smart!"

My mother thought I was smart because I had the sensibility to do as she said—to go back to my room and entertain myself so she could rest, even though I was only two. For her it is a story about my prodigious independence. For me it is a story of what never happened—connection, warmth, and nurturing attention between mother and child.

I still have that photocopy folded in thirds and tucked into one of my own journals. I guard it like a keepsake. It is an artifact of how my mother and I communicate with each other—through words on the page more so than words spoken out loud. It is a snapshot of a childhood of omission that affected my nervous system for life.

I had a childhood . . .

I don't remember being the two-year-old my mother described in her journal, but I do remember sitting in silence. Our family lived in a red brick house on a hill, on a dead-end street with a smattering of far-apart homes. To take care of the house, my parents hired a woman named Lydia Servano, whose name is eerily like my mother's, Lydia Soriano. And so with my father working, my mother also working or resting behind a locked door, Lydia cooking or cleaning, and my brothers out riding bikes with the neighborhood boys, I was left to wander through the world within our home, alone.

When I was three or four, one of my favorite places to pass time was underneath my parents' baby grand. I squatted underneath its elegant black body, tracing its wooden slats left and right, up and down. I felt comfortable there, invisible and protected, with the shelter of ebony wood and brass looming strong above my head.

From underneath the piano I would look around the room, which I thought of as the world outside. It was a beautiful room with cream carpeting and a long bay window with houseplants flourishing even in winter light. The plants were so lovingly cared for by both Lydia and my mother. There was a strawberry-drop coleus, three leggy spider plants, a wax-green calamansi, and a hibiscus with candy-pink blooms. The flowers seemed to look at me and preen.

Other times I wandered around the house, keeping company with objects and playing with friends who were things. I used to talk to these objects in my head, greeting them as if reciting an inverse version of *Goodnight Moon*.

"Hello, Brains," I said to the photo albums of my father's brain surgeries.

"Hello, Ruby Red," I said to the single red boxing glove autographed by "The Warrior," Evander Holyfield. The glove floated in a sealed acrylic case that I once tried to crack open with a steak knife.

"Hello, Little Gun," I said to my father's Smith & Wesson .38 revolver that lived on a shelf in his closet—a high shelf, but not too high for me to reach while standing on a chair. I loved its heft and how perfectly it fit in the palm of my hand.

"Hello, Pretty Place," I said to an oil painting of woods and a stream. The stream led to a clearing of light and to somewhere, I thought, that must be beautiful.

A fearful childhood . . .

After I was born, my mother had her tubes tied. I don't remember how I learned this. But I do remember wondering whether she had wanted them tied after my second brother was born.

When my mother wasn't resting behind a closed and locked door, she allowed me to play at her feet while she did paperwork for my father's business or while she put on makeup and curled her hair. All the while, we never made eye contact. We only spoke when I ventured a question, "What are all those papers?" or "How does a curling iron work?" to which she'd give half-hearted and absent-minded replies. Children learn to pick up on these kinds of subtle cues. Because of this inattention and because she often rebuffed me when exhausted, I grew up believing that my mother hadn't really wanted a third child.

My father had once said that he convinced my mother to have another kid because he knew it would be a girl. When I came out of my mother's womb with my own vagina, I imagine she was relieved to see that my father's prophecy had come true. By the time I was four, I had all the trappings of a spoiled princess: a closet full of dresses, a violet-carpeted room, and flowered wallpaper with matching curtains. This was one way my mother expressed care for me—through ruffled clothing and interior decor. In this showcase of a bedroom, I spent time with books, which were my friends, and my stuffed animals, which were not.

The stuffed animal I remember most was named Lily, a tiger with shell-shocked bottle green eyes whom my brother had won at a fair. Lily sat on her haunches with her two front legs extended like a zombie. She was hard and hollow, and her fur was glossy but more plastic than soft, like Astroturf. She didn't have a cuddle in her body. She always looked ready to pounce.

I was convinced that Lily would come to life in the middle of the night to attack me. So I thought I should attack first: a Chinese star from my brothers' room could slash her open, or one lit match could eat her alive. But if I hurt her, the other stuffed animals might come to exact revenge. So I settled on keeping her where I could see her. I lined her up on my dresser next to the other suspects. I kissed each and every one of them good night in an attempt to pacify them, but gave Lily three kisses for extra security, then went to bed with the covers pulled up to my eyes.

Lily was one example of the vast interior life I led, partly because my parents kept me sheltered at home, and also because it was my nature to be sensitive and introspective. In the absence of much contact with anyone outside school and family, I was left to have an intimate relationship with my imagination. There were wonderful sides to this. I found a turtle shell in the backyard and imagined my extended family as a village of turtles, inhabiting the nooks and crannies of the preserve behind our house. But there were also fearful, lonely sides that believed Lily and the rest of my stuffed animals would harm me if I didn't act right.

Lily was not the only thing I feared. I was scared of everything that went bump in the night: the creaks in the attic, the wailing wind that pushed its way under our doors, the oil painting of a crying clown hanging on my bedroom wall.

But what I feared most was that I was a burden and that I was never meant to be here at all.

Once when I was six or seven, I watched a scary movie with my brothers that was inappropriate for my age. After a scene where a devil emerged from a pit of blood, I ran out of the room to find my mother.

"I just saw something really scary, Mom!" I cried.

She looked over at me with a scowl, her arms raised in midair, folding a towel, and my first thought was that I should not have interrupted her.

She snapped, "Well, that's your fault!" and turned away.

Her words hit my still-small body like ice water. For such a petite woman, she had the force of a polar bear when angry. I wanted to dissolve and disappear. Then a wave of regret washed over me. Why had I bothered her? Why didn't I just keep quiet, suck it up, and act brave? Who was I to come to her with my ridiculous needs?

A few days later, I sat on my mother's bathroom floor while she dyed her hair. Despite her snubs, I still longed to be close to her. I watched her wield her dyeing tool: a new toothbrush, which she held delicately between fingers tipped by perfectly filed nails. She dipped the brush into a bowl of water mixed with powder from the Asian store, and its bristles turned from white to charcoal black. With gentle strokes she brushed the dye through the wisps of gray at her temples until they blended in with the rest of her short-cut black hair.

I would later realize that my mother was as manually skilled as my father, but instead of using scalpels and drills, she used brushes and scissors and whatever else she could find in the cabinets. She was like a Filipina MacGyver, constructing clever solutions to household problems. She once rigged up an automatic plant-watering system with surgical tubing, binder clips, and plastic gallon milk bottles cut in half.

As she continued with the ritual of dyeing her hair, my mother began to tell me a story of when I was two years old and she had a headache. It was the same story she later photocopied from her journal and mailed to me. My mother rarely addressed me so directly, so I made it a point to listen. Looking back, this was one of the only ways my mother employed what child development specialists call "narrative elaboration," or the process of storytelling to help construct meaning in a child's life.

I constructed meaning from the story right away: if you want to be a good kid, deal with the hard things yourself. Don't be a burden, swallow your fears, and leave your mother to rest, to sew, to dye her hair, to engineer household systems and take care of her plants. In other words, control yourself and leave your mother be. Wasn't I such a smart child?

A playful childhood . . .

The levity I remember from childhood came almost entirely from playing with my brothers. They were four and six years older than me, so I spent a lot of time alone until I grew old enough to keep up with them. By the time I was seven, we were our own little clan of adventurers, browned by summer sun and made red from winter wind.

My brothers and I would leap from the retaining wall beside the house to catch a rope knotted to a tree branch, then swing one story down to the ground. We would ride bikes down our steep driveway, hurtling across the street until we landed in a ditch. These moments still make me smile. Physicality was an important form of emotional bonding for me. Playing with my brothers connected me to something outside of myself that made me happy, because as much as I was an introspective child, I also wanted to be part of the world.

Unsupervised as we were, my brothers and I did things that would give helicopter parents heart attacks. We hurled knives into trees, shot BB guns into the air, torched anthills and bushes with matches and bike grease, threw bricks and balisong and Chinese stars, played with our father's gun and loaded the barrel with bullets. Thankfully, my brothers refrained from cocking the hammer or pulling the trigger. To this day I think it's a small miracle we survived.

A normal childhood . . .

In my twenties, I had a dream about my mother. She was wrapped in gossamer and silk; as I approached her, I saw she was ensconced in a cocoon. Waking up from the dream, I realized that this is how I saw my mother: bundled up and protected. She lived her life catering to my father's every need while my father, in return, catered to hers. She set up systems to keep the house organized and admits she always liked things malinis, a Tagalog world which roughly means "immaculate." She used Erlenmeyer flasks and beakers, ordered direct from a chemistry lab, to dispense juice and iced tea. As a trained industrial pharmacist, she knew the laboratory glassware was easy to sterilize. She was vigilant about protecting us from germs. She didn't allow me to crawl because she wanted my hands and knees to stay clean.

I wondered if she ever wanted to emerge from that cocoon of protection into a different life, where she was a different person, a carefree writer perhaps, someone with two rather than three children or maybe no children at all, an artist who traveled the world oblivious to germs, an accomplished scientist and pharmacist, or a nun. Or perhaps she was most content there in that cocoon, limited, protected, and sheltered, like the childhood she offered me.

In the mornings in our red brick house on that isolated hill, my mother would open the blinds, peer out the window, and mutter, "They can see us." Then she would snap the vertical slats shut. I would ask her whom she was talking about, and she would reply, "The neighbors!" Our nearest neighbors were shielded behind trees, and the neighbors behind our house lived all the way down at the end of the cul-de-sac in a dark area near the surrounding preserve. We never saw them, and yet my mother made sure that the blinds, as well as our six-foot-tall iron gate, stayed closed. When solicitors walked through the gap between the fence and the gate, she sent them away with a frown and a scolding. She reserved a special disdain for Girl Scouts, whom she believed peddled poisoned cookies. By the evenings, my mother was back behind her closed bedroom door. At night I read myself bedtime stories, and when I found spiders under my pillow or wanted to be tucked in, it was one of my brothers who would come to my aid.

For a long time, I thought it was normal to be so guarded and afraid. I thought it was normal to wake up every morning feeling sad and anxious and for silence to seal one's mouth for days.

I thought it was normal to expect to die young and to go to sleep every night believing you might not wake up the next day.

I thought it was a natural rule that if you fall, you keep going. You keep going even if you've snapped your ankle or broken your tailbone. You keep going after the falls because there is no one there to see you and no one there to pick you up.

This all seemed normal because I had nothing to compare it to—and because it's impossible to want butterflies when all you've known is a cocoon.

Then I became a teenager . . .

By the time I was ten, my brothers were at East Coast boarding schools, which was part of my parents' plan to achieve the American Dream. Our adventure clan disbanded. Even though as a preteen I longed to spend more time with my peers, I continued to spend most of my time alone in our house on a hill.

Though my mother took me to violin class and gymnastics, what I wanted most was interaction rather than activity. My strongest memory of interacting with anyone in this era was a moment with my mother, when I got my first period at eleven years old. She said some version of "You're a woman now," and the next day handed me a copy of *Our Bodies, Ourselves*. At the time I didn't know how progressive the book was, and we never talked about it. I wish we had, because the book was a missed connection point for us, a springboard from which we could have discussed our common anatomy through a scientific lens. Instead of forming a bond through biology, we grew further apart because of it.

When I was thirteen, my parents allowed me to go to my friend Kay's birthday party. I was frantic with excitement. Because my parents did not allow me to hang out with boys outside of school, I had long been missing out on ice cream and burger outings, park trips, and other after-school socializing. This was my chance to win back friendships lost from my absence.

Unbeknownst to my parents, Kay had invited our whole class over to her house. Despite my parents' worst fears, the party was good clean fun—just pizza and pop and talking and music and a movie. The boys were mostly gone when my mother picked me up, except for one kid named Dave who came trotting down the stairs just as my mom came in the front door. She drove me home in silence. But when my father came home late that night, she asked me downstairs for a talk.

We rarely had familial talks, so I knew this couldn't be good. My mother crossed her arms. Smoke seemed to curl out of her ears. My father verbally fumed: "What were you doing there with boys at that party? Don't you know what could happen? Are you being stupid?!"

A dam broke inside me then, and I went from silently acquiescing to screaming. I was starting to see what was actually normal. While my brothers had been allowed to roam free and Kay and my other friends were out forming deeper bonds, my parents kept me wrapped in a cocoon or locked in a cupboard, dusted and protected like bone china.

"I hate you!" I yelled. "Why do my brothers get to do all this stuff and I can't? Just because I'm a girl? What do you think, I want to get pregnant? You don't trust me. It's so unfair!"

They accused me of being ungrateful, my father threatened me with the belt or the slipper. The irony is that months later, they did allow me to go to a coed boarding school. They must have thought there would be enough structure and security to keep me out of trouble. They didn't have to worry. By then I'd internalized the strict parent within, and besides a few close friends, I kept mostly to myself.

And then I went to therapy . . .

I am twenty-six years old and lying on a mat on the floor. More than a decade after leaving my parents' home for high school, my anxiety and pain have gotten so bad, I've decided to seek professional help. This is my sixth session with Bret Lyon, a therapist I found through an article in a grocery store health magazine. Bret has supplied me with a fat brown teddy bear that I now squeeze between my crossed arms. I'm sobbing and feeling bad about soaking the bear and the lovely red ribbon tied about its throat. I wonder for a moment if the bear can breathe with me crushing it so hard against my chest. Bret sits next to me in a chair and tells me it's okay to cry and that I'm not alone. When it feels like all my body fluids have drained through my eyes, I am able to get up and sit on a sofa to face him.

I try to take a deep breath. Air moves from my stomach to my lungs, then catches in my throat. A wave of nausea overwhelms me, and I begin to sob again. I feel like I am throwing up grief.

"It's okay," Bret says. "Let it out. You're here. I'm here. Can you feel the chair under you and feel that it's supporting you?"

He hands me a new box of tissues, which I manage to grab between arm flails. I dutifully nod in response, a twitching nod, thrown off by my shaking shoulders. When the waterfall of tears becomes a trickle, he speaks.

"You have an attachment wound," he pronounces. This is the first time he has uttered what sounds like a diagnosis.

I begin to cry again and proceed to make my way through every tissue in the second box.

"You were emotionally neglected and were never taught how to depend on a caregiver," he says. "Forgive yourself. This is a big deal to work through."

He waits for me to blow my nose, then continues, "A child is supposed to be able to move back and forth from caregiver to independence. This is why toddlers will walk away but always look back. Then when they return and touch their caregiver's leg, they feel the safety to try to venture out again, and they do. This back-and-forth is key to developing a baseline of security."

I quiet my sobs to listen more closely. I picture this back-and-forth movement in my head, trying to match it to a memory from my own childhood.

"When we don't get this baseline of security," Bret goes on, "we know even as very young children that we are missing something that we deserve. And we begin to grieve it even then, but may not fully face it, even as adults, because it can be so painful."

This is my cue to start outright crying again as I realize that the back-and-forth he is describing is difficult to recall because it never happened. It is the first time I'm able to imagine the omissions of my childhood. Grief washes over me like a tsunami, and I think my heart is about to drown. I don't yet feel anger. That will take years to come. Instead, I feel something akin to shame, the shame of being deeply seen when you don't feel like you deserve a second glance.

"Be easy on yourself," Bret says. "Attachment wounds run deep. They can scab over, but they don't fully heal."

"Well, what do you think I did wrong?" I ask in what comes out like a whimper. "Why didn't my parents give me that back-and-forth?"

His face softens from straightforward and declarative to a kind and sympathetic sort of pity. Shame washes over me again.

"It's common for people who've experienced neglect to think it was their fault," he says. "But it wasn't your fault at all. Your parents just didn't have the capacity to provide you with a secure attachment. Young children will shame themselves before they blame their parents because they need to believe their parents are faultless in order to survive."

So, this is an actual thing, I think to myself. *It's not just my own sad little shortcoming.* I feel as if I've broken the surface of the ocean and finally come up for air.

"But shouldn't I have gotten over this by now?" I say out loud.

He scoffs in a way that makes my body relax. "You're just finding out about it! Most people who experience emotional neglect wrestle with the wound their entire lives."

I slump back on the sofa, a mirror of the soggy bear I left slumped on the mat on the floor. I had thought of my childhood as normal and even happy, something that I later learned is common for emotionally neglected children. But at the same time, I knew something was wrong, and I thought that something was me.

Bret says matter-of-factly, "Because of this work you are doing, the frequency of episodes will be less and less. But once in a while, you will still have to retreat to your cave to lick your wounds."

And learned about being unlicked . . .

In the mid-2000s, scientists at McGill University conducted an experiment with rat parents, some of whom licked their pups a lot, and others who licked their pups rarely or not at all. They found that the rat pups who were licked often grew up calm and collected. In contrast, the rat pups who were licked infrequently or not at all grew up skittish and easily startled. Unlicked rats were nervous rats.

What exactly caused these differences? It wasn't genetics; it's not that the pups were inheriting a calm gene from their high-licking parent or a nervous gene from their low-licking parent. It was the actual act of licking, or omission thereof, that changed the biochemistry of the rat pups' nervous systems. This difference in nurturing during infancy determined whether or not the pups would have a regulated response to stress. What's more, these epigenetic, or above the gene, modifications became lasting changes that could affect the rats' behavior throughout their lives.

When I learned about this experiment, I finally believed what Bret told me about the need to lick childhood wounds through adulthood. I found comfort in knowing there was a biological basis to my outsize skittishness and fears, at least if I compared myself to a rat.

Similar changes are thought to occur in humans who have experienced neglect. As infants and children, our developing nervous systems rely on responsive interaction with a caregiver—vocalization, facial mirroring, and nurturing touch are essentially the human equivalent of rats licking their young.

In his famous "still face" experiment, scientist Ed Tronick studied the effects of mothers' facial interactions on their babies. In a video that documents one of these interactions, a mother at first plays with her one-year-old baby. The baby makes a noise, and the mother makes the same noise back. The baby scrunches its face and reaches for her, and she scrunches her own face and reaches back. The one-year-old points, the mom looks and smiles, and the baby bursts into gurgling laughter.

But when the mother puts on a still face, the effect on the baby is astonishing. The baby reaches out again and coos, but this time, the mom keeps her face flat, unresponsive, gazing at an unspecified point rather than directly at the child. The baby quickly notices and does a double take as if to ask, *Wait, what happened?* Then the baby proceeds to do everything they did before that had been successful in getting the mother's attention. Within two minutes, with no response from the mother, the baby grows distressed and begins to cry, sniffling at first, then sobbing. It's heartbreaking to see the baby's misery. The baby attempts to turn away and escape from the high chair. At last the mother says, "Okay!" and resumes playing with the child, at which point the baby immediately smiles and settles back down.

This process of interaction is something child development specialists call "serve and return," which is essentially what Bret Lyon described as "back-and-forth." When this reciprocal interaction is absent, then, according to Dr. Megan Gunnar, "you are literally pulling away the essential ingredient of the development of human brain architecture."

Watching this video, I felt a wave of compassion—for the baby, and for the child version of me. This may have been the first time I felt empathy for little Jen. I knew that my emotions had never been small, slightly needful things. My emotions were always big and voracious and unruly. Now I understood that they needed a larger and wiser body to help process them, and that I needed this larger and wiser body to help me establish a baseline of calm.

In our home there were no "Hi, how are you"s or "How was your day?"s, much less attentive comfort when distressed. I mourn the loss of everyday language—in any tongue—that would have shown that we cared about one another, that we saw one another, that we understood. Instead of the easy flow of conversation and reciprocal engagement, I recall silence and the sensation of my body made rigid. Since interaction is key to understanding where the boundaries of our bodies end and the rest of the world begins, I had no idea where my body stood in relationship to others. I only knew that I had to brace myself against everything.

When caretaker interaction is absent on a regular basis, this omission tells the child that the world is an unpredictable and unsafe place, which chronically activates a child's stress response to help them survive. This can predispose "unlicked" babies to chronic disease later in life, including diabetes and heart disease, as well as mental illness and substance use. It can also be a cause, or at least a predisposing factor, for chronic pain, PTSD, and complex PTSD. Without the regulating comfort of a caregiver, we become wired for stress and fear, and we can start to believe that everything is a threat, even a green-eyed stuffed tiger named Lily.

Tronick points out that babies can recover from momentary inattention. But if, after just a few minutes of nonresponsiveness from a caregiver, babies get intensely distressed, imagine what happens when "still face" is what a child receives from caregivers most, if not all, of the time.

And then I kept digging . . .

Psychologist Enod Grey calls neglect the "silent abuser." Others have called it "abuse by omission." While there may be concrete signs in cases of physical abuse and severe physical neglect, identifying emotional neglect is like trying to hold water in your hands.

I was well-fed, bathed, taken to the dentist and pediatric checkups, and sent to good schools. I was able to develop a sense of healthy interaction with other children through my relationship with my brothers. And I now know that my mother made sacrifices to care for me in her own way: driving me to school and music lessons, helping me with dioramas and science projects, and sewing me handmade Halloween costumes even though I was never allowed to trick-or-treat. She cared for me like an operations manager: materially, and through logistical and project-based support.

But I experienced emotional neglect involving "chronic understimulation," or the lack of intimate serve-and-return. I also experienced physical neglect, such as failure to crawl and failure to roll over—both consequences of my mother trying to keep me safe from germs—as well as failure to attend to physical injuries. Later in life, I learned that this form of neglect prevented me from reaching neuromuscular developmental milestones and likely contributed to mild scoliosis and both peripheral and centralized pain.

No child gets everything they need from their caregivers, but neglect is a deeper level of denying a child's most fundamental needs. Still, it's easy to dismiss and even mock neglect, especially emotional neglect. In certain social circles, we compare latchkey kid stories, create a hierarchy of hard-knock childhoods, and laugh about who had it worse. In larger society, patriarchy and misogyny teach us to believe that having emotional needs is a sign of weakness. Getting them met is an indicator of a spoiled child.

But science shows us what healthy people already understand: that emotional needs are on par with material needs, and that emotions are powerful enough to affect both our psychological and physical health for life.

A river of neglect . . .

When my mother was about to turn eighty, my brothers and I tried to organize a big party for her. "Let's go on a boat like we did for your fiftieth anniversary!" my oldest brother suggested. "Let's have a brunch or dinner with music," my second brother suggested. "Let's have a party in Las Vegas and invite all your friends," I said. She didn't seem enthusiastic about any of our plans, and so, exasperated, I finally demanded to know, "So what is it you want?"

Her face crumpled. She went from looking distant and reserved to soft and defeated. "I don't know what I want," she said quietly through a river of tears. My shoulders, which had been raised like a cat's hackles, suddenly fell in an act of surrender. I got up from the chair where I had been facing my mother and brought her a box of tissues. "It's okay, Mom, we'll plan something and hopefully you'll just have fun."

That moment was uncomfortable, and it was also enlightening. I felt compassion for my mom. How could she have addressed my emotions when she was unable to attune to her own? This moment left me to wonder what feelings she has had to suppress in her life and why. What has she had to carry in her body, and what was the source of her own "low-licking" behaviors which then affected me?

Hurt and hope flow downstream . . .

Years before he died, my dad admitted that he knew he and my mom kept me alone a lot. My mother refrained from commenting. The three of us were sitting at their kitchen table in Las Vegas, and my father was waxing poetic about his neurosurgery career. As an aside, he mentioned that I was left alone because there was so much work for both him and "Mommy" to do. He sounded repentant. I didn't interrupt him. I simply nodded and allowed the comment to slide back into the current of his reminiscence and the steady stream of my mother's silence.

Though I never acknowledged it, this small recognition of their neglect meant the world to me. For all my parents' desires to protect their little girl from danger, it was in part their combination of neglect and control that left me most vulnerable to harm in the form of mental illness and chronic disease.

In the process of my own healing, I have come to treasure the self-reliance I learned from neglect while rejecting the imposed silence. I have become a butterfly catcher, a seeker of elusive recollections. I cast my net around belly and breath, and when I catch a rare specimen, I pin it to the page with my words.

One of these butterflies escaped from my mother's mouth when my own baby was turning into a toddler. I was talking about what a strong temperament my son had when she interrupted me and said suddenly, "Don't neglect him!"

I cocked my head at her and asked, "Why would you say that?"

"I don't know, just be sure not to neglect him," she said, then turned away.

I thought this must be her subconscious talking. But if there is a part of her that knows she neglected me, she won't tell me directly like my father did. It's simply not her style.

I have continued to learn things about my mother indirectly, through words on the page. On a recent trip to Las Vegas, I was doing laundry in her house when a pile of papers slipped off a shelf onto the floor. As I gathered them together, my own name caught my eye. It was scrawled on a sheet of yellow legal paper followed by a few paragraphs in my mother's handwriting. I narrowed my eyes to read through several crossed-out words. It began: "They say that girls are sugar and spice and everything nice, but that is bound to set you up for disappointment." I smirked and read on: "I thought that having a girl would be like having a doll that you could dress up, that would listen and smile and go to sleep when told. But Jenny throws tantrums and has always had a mind of her own."

As cutting as these words could have been, when I read them I laughed out loud because they explained so much. I apparently wasn't always as quiet as I had remembered myself to be, which matched the copious volume of emotion I knew I held as an adult. And I finally had the answer to my deepest question of whether my mother had wanted a third child: she had indeed wanted a girl—just not the girl she got.

I wonder now if it would make sense to her or if it would confound her more if I explained that I don't even identify as strictly a girl but as nonbinary. A part of me draws some delight from imagining another layer of myself that she wouldn't quite be able to see.

I have come to feel gratitude for what my mother was able to give me. She worked hard to provide, she drove me and my brothers to activities and a faraway private school, she made sure we ate healthy meals and stayed hydrated and worried constantly about our well-being. Many children receive far less, and so for what I have received I am grateful. I no longer expect any more. I realize that caregivers can only give more when they have had the opportunity to face their own traumas. And Bret had helped me realize that my parents gave me just enough resourcing that, when it has mattered most, I've been able to make the right choices to heal.

But I still long for the closeness and comfort that nurturing attention would have given me when I was small. And as I struggle to parent my own child with more emotional connection than I received, I commit to having conversations that never happened, that I wish a caregiver would have had with me:

Yes, it must be scary to think of your stuffed animals as monsters who might wake up to hurt you at night. I promise it won't happen, and I'm here to protect you.

Yes, it must be frightening to believe you could die at any moment. You're safe here though, and I'll stay with you until you fall asleep.

Mobility

I.

I emerge from the archives of the main library at the University of the Philippines, driven by a desire to see the world beyond microfiche and books. Blinking into the westward sun, I stand at the top of the library's grand steps, unsure of which way to go.

It is the summer of 1997, I am twenty-one years old, and I have come to Manila on a grant to do research on American family planning programs in the Philippines. I need a break, so I descend the stairs to meander across the manicured grass of the Academic Oval, a stretch of green the length of ten football fields. Pain crawls along my arms. Paging through books and taking notes has aggravated the normal ache in my wrists, and so I shake them out to numb the pain. As I walk and stretch my limbs, I try to imagine my parents as students here some forty years before.

I pass the engineering hall and picture their lives on this campus as a series of near misses with each other. My mother studied industrial pharmacy, and my father was premed and went on to UP's medical school. But they didn't meet until they had both graduated and moved on to America.

I envision my mother at twenty years old, walking across the oval in a petite, cinch-waisted dress, her black hair curled and pinned. She was the class salutatorian, a math whiz and natural student who also loved to socialize. She had a wide circle of friends, and I imagine she knew how to turn on her electric smile when greeting young men in tucked shirts and khaki slacks at one of the monthly mixers between engineering and pharmacy students. Meanwhile, my father would be studying.

Behind the engineering building is a Quonset hut, a semicircular structure of stainless steel. The American military installed dozens of these at the outbreak of World War II, but after the devastation of Manila at the end of the war, only a few remained. This is where my father lived as a premed student because it was the cheapest housing on campus. The hut had an open plan with no interior walls, so my father and his fellow dormmates—a decommissioned army colonel, some engineering students, and a bully—had to move lockers to form partitions between their sleeping areas. My father's grandfather, a master carpenter, built him a desk because otherwise there were none. I picture my father hunching over this prized piece of furniture, burning the midnight oil, his ironed shirt untucked and his black horn-rimmed glasses slipping down to the tip of his nose.

I continue to wander the UP campus even as my ankles tremble from the walk. At the opposite end of the oval, I reach the Oblation, a statue that is the most well-known symbol of UP and features prominently in my parents' alumni swag. It is a sculpture of a naked person cast in bronze, brown and dignified, arms splayed, chest thrust high, and head thrown back. Their genitalia, originally exposed by the artist, was covered by a bronze fig leaf four years after the sculpture's unveiling. The statue stands "decent" but proud, 3.5 meters tall to symbolize the 3.5 centuries of Spanish colonization that taught Filipinos to revere oblations to God. But rather than an offering to God, the artist created this sculpture as a tribute to the Filipino offering themselves up to the nation.

The Oblation was installed facing west in tribute to the university's American origins, which begs the question of which nation the Oblation is offering their body to. In my mind, the statue's pose suggests an imminent leap from the pedestal of rocks, representing the Philippine archipelago, and into the ocean of grass and concrete that lies before them. The Oblation seems ready to swim through Manila Bay across the sea.

II.

Before my parents crossed the Pacific Ocean, they were born on the Philippine Islands in the 1930s, which means they were born American colonial subjects in the Greater United States. The "Greater United

States" is a term that historian Daniel Immerwahr uses to name America, its former colonies, and its current territories. It is a name that reflects what the US has worked hard to obscure: American imperial control of six island nations and enduring influence over their economies, as well as the cultures and psychologies of their people.

My father grew up in Tondo, Manila, a neighborhood you could compare to the South Bronx in terms of poverty and crime, as well as grit and hustle. It is also a barrio known for breeding revolutionaries. Andrés Bonifacio, who led the Katipunero revolt against Spanish rule; Marina Dizon, Katipunera and keeper of rebel documents; and Macario Sakay, who continued the rebellion against American forces after Emilio Aguinaldo surrendered, were all from Tondo. But my father was more enamored of American role models like Sugar Ray Robinson and Frank Sinatra than he was of the revolutionary figures from his hometown.

From the late '30s through the 1950s, Tondo was what my father called a "dignified but rough-and-tumble working-class neighborhood." After World War II, Tondo's population surged as people moved from rural areas in search of jobs made scarce by the postwar neocolonial economy. Poverty rose, as did the threat of violence from street gangs.

Despite this, my father describes his childhood as humble but happy. He was surrounded by cousins, including his best friend and first cousin, Pepito, with whom he played street games like sipa, patintero, and trumpo. He and his cousins swam in the bay and in the tributary of the Pasig River that ran near the tree-lined stretch of Plaza Moriones. In the afternoons he went door-to-door for his uncle, the town doctor, to collect coins from patients who were slowly but surely paying their bills. On weekends he sold lottery tickets and sampaguita wreaths at Plaza Miranda in front of Quiapo Church. But what he most loved was accompanying his mother to the Divisoria fish market, where he would sit on top of milk crates and read rented American comics. His favorites were Batman, Superman, and the Green Lantern.

My mother also recalls her childhood as happy. She was born and raised in Bocaue, Bulacan, a province connected to the city by General Douglas MacArthur Highway. Though her own mother was raised poor and dropped out of school to sell food at the sabong, her father came from an established family of teachers who worked in the American colonial education system. My mother had a comfortable upbringing because her mother was determined to provide all the advantages she never had. My mother took piano lessons and went to private Catholic school and finishing school. She loved being surrounded by her mother and brothers, as well as a network of extended family members who always looked out for her.

When my parents took my brothers and I back to the Philippines, we never spent much time in the barrios where they grew up. Instead, we stayed at the Manila Hotel and remained sealed inside air-conditioned cars with tinted windows. Once, at my mother's ancestral home in Bocaue, my brothers and I went walking on the shoulder of MacArthur Highway, just to see what was around town. We were about to flag down a jeepney when a van pulled up swiftly beside us and flung open its door. *We're about to get kidnapped!* I thought. But it was only my cousin coming to scoop us up. Our parents had sent him to fetch us so we wouldn't get kidnapped by somebody else.

My parents didn't seem to want us kids to really touch the Philippines. Perhaps they believed it was no longer good enough for us since they had transplanted themselves in supposedly superior American soil.

III.

Back on the UP campus, I've walked a long way in the heat, and my skin is covered in a film of sweat. My scoliosis and neck compression flared in the library, but the heat eases my pain. At this point in my life, I have seen two doctors and one counselor, and from them I got one X-ray, some candy, and a bottle of Aleve. My only other pain remedies are movement, heat, and distraction. Academics are a distraction, and so I stop at a stone table under an acacia tree and pull out my notes for review.

I'd found through my research that in the 1970s, many elites, both American and Filipino, believed that family planning programs were necessary to achieve upward mobility for the Filipino masses. Deteriorating conditions in Tondo motivated the effort. Many Filipinas did indeed want access to birth control—and still do. But in practice, the quota-based American-controlled programs contributed more to coercive population control than they did to upward mobility and reproductive freedom. The main question I was asking through my research was: Mobility for who?

I look up from the table, wishing I could discuss these findings with someone. It's my first time back in the islands without my parents. Since I've come on a college mission to complete my undergraduate thesis, they are okay with my being here alone. Their desire for my academic achievement trumps their normal fears that I will be abducted and held for ransom. Plus, they are relieved that I am continuing my studies at all. The previous year I had almost become a college dropout, and even worse in their eyes, a Harvard dropout.

The people around me at Harvard were royalty—the children of European rulers and Middle Eastern officials, as well as New England blue bloods with last names etched on plaques across campus. I couldn't relate to their luxurious lifestyles of yachting over long weekends and eating out every night, nor could I understand their breezy attitude toward education: "Just go get that Ivy League B, then blow it all away," one of them said to me, sniffing and tweaking her nose with her forefinger. At Harvard I was a fish out of water, even among other Filipino American students who stood on the sidelines during campus protests for ethnic studies and who even joked about organizing a counter-protest for "assimilation studies." Looking back, I felt suffocated by New England colonial culture and my fellow Filipino American students' colonial mentality.

In the spring of sophomore year, I took a semester off to go west to the California Bay Area. There, the warmer weather, slower pace of life, racial diversity, and progressive politics all mixed together to form sweeter water I longed to swim in. Every cell in my body wanted to stay in this fluid environment rather than retract to the rigid culture of elite competition that dominated my Harvard world.

The only thing that made me return was the thought of the shame it would bring to my parents if I didn't finish school. My mother was the first woman in her family to go to college, and my father was the first in his family to get an advanced degree. How could I break this lineage of advancement by becoming a college dropout?

And so here I am, more than eight thousand miles away from Harvard at my parents' alma mater, trying to complete my last graduation requirement. I am grateful for the resources that have brought me back to my parents' home. After all, this thesis research is really an excuse to get a different type of education, one of being immersed in the place where my family first took root.

A stream of students pours out from the nearby building, and a group of them pile around the table next to me. I look over, but they pay me no mind. There are nearly two dozen kids, of all seeming genders, and they're leaning down to look at something on the table. I crane my neck and see that it's a tiny white puppy with a purple bow in its hair. One of the girls says something in English that I can't quite make out, and they all laugh. Arms draped around shoulders, hips jostling against hips, these students are an easygoing, amoeba-like-being made of many connected bodies, all petting one dog.

I refocus on my notes while also trying to scare up the courage to talk to them. Completing this thesis is the last leg of the educational race to the top in which my parents have so doggedly placed me. But as I sit alone at the table, with my body touching only the rough bench below me, I can't help but think, *Mobility to what end?*

IV.

My mother used to commute from Bocaue to Manila for school. Her family could afford to employ a driver, and so they hired a good one who would go "very, very fast." Meanwhile, my father spent summers in Bicol working at his grandfather's shoe business and tannery. He rode the night train from Tondo to the southeastern reaches of Luzon, a journey of more than fourteen hours.

My parents traveled a distance to attend UP and were both at the top of their classes. My mother graduated in five years with a degree in industrial pharmacy, and my father graduated in eight with a degree in general medicine. Their mobility seemed linear. They were on track to achieve great things.

Since my mother in particular was good at math, she would be able to vouch for these facts: the shortest path from one point to another is a straight line. But only if you live on a flat Earth. On curved terrain, the shortest distance from one point to another is actually a great circle.

V.

My stomach rumbles, and I think I should go find some lunch. I'm not sure where to look because I've been unable to muster the courage to ask one of the students for help. I can't bear the thought of all those eyes turning toward me. Although I'm thirsty for connection, in the end, I prefer to remain unseen.

I get up from the table and start to walk away. As I do, I hear one of them say, "Bakit siya mag-isa?" Which literally means "Why are they just one?" I stand out to them not because of my clothes or my short hair or the way that I walk, but because I am alone.

A jeepney putters by, and I follow it with a longing gaze; the ubiquitous mode of Philippine mass transit is one I have admired for a while but from afar. This particular jeepney is dull-gray, small, and, above all, functional. The rectangular sign affixed to its front says, "Ikot," scrawled in black letters. "Ikot" is the Tagalog word for going around in circles.

These Ikot jeepneys transport students around the oval formed by Osmeña and Roxas Avenues. How appropriate, I think, that these roads bear the names of two successive Philippine presidents—Osmeña, who was the last to govern the US colony, and Roxas, who was the first to govern the neo-colony. The two streets join to move you not from one point to the next but in loops, back to where you began.

As I watch the Ikot jeepney round the bend, I imagine what my life would have been like if my parents had stayed in the Philippines and I had been born here. Would it be a dead end, like my father predicted? He thought there was little prospect of upward mobility in the Philippines. It was too much of a corrupt crony system, and neither he nor my mother were in the circle.

With envy I watch students crossing campus, walking with their hands in each other's back pockets or arms curved around each other's hips. I think of my relatives who have stayed: my kuya, who coordinates transportation for the government, and his wife, who runs a school bus service for K–12 kids. I think of my ate, who works for a catering company and lives in my mother's ancestral home, fosters street dogs, and spends holidays traveling across the islands in a Tamaraw Jeep. I think of my niece, who is a dentist, and my nephew, who followed in my mother's footsteps and is now a practicing pharmacist. My niece and nephew often commute together from their home to my niece's dental office in Makati. The family regularly travels across Metro Manila to gather at our ancestral home in Bocaue, to eat and tsismis and sing karaoke. They share a seemingly endless spate of inside jokes.

Would it have been so bad?

VI.

Unlike poorer migrants with few prospects in the Philippines, my parents had the privilege to stay or go. After graduating from UP, my mother opened a pharmacy in Caloocan City. Starting on a shoestring budget, she rolled over her profits until it was fully stocked and gave away drug samples to people in need, unlike some of her colleagues who sold them for profit. For his part, my father was offered a position at Philippine General Hospital. His aunts were also planning to build a clinic where he could become the barrio doctor. Both my mother and father were poised to devote their newly acquired professional skills to communities in the Philippines.

But America had other plans.

In 1898, when Americans bought the Philippines from Spain for twenty million dollars, they at first didn't know what to do with their purchase. William McKinley famously got down on his knees and prayed for guidance. In a speech that epitomized "The White Man's Burden," which Kipling wrote specifically about the Philippines, McKinley admits that he didn't want the Philippines at all, but since it came "as a gift from the gods," he decided his only option was "to educate the Filipinos, and uplift and civilize and Christianize them." Thus McKinley set in motion the narrative of America as benevolent empire, who, unlike Spain, would deign to school their "little brown brothers" and shape them in the very image of American light.

The University of the Philippines was a key vehicle in this enlightenment project. Launched in 1908 by colonial Secretary of Public Instruction William Morgan Shuster, UP was modeled after American schools like the state universities of California. By midcentury, after the

Second World War, it became a feeder for selective migration to the United States.

In the 1950s and early '60s, American law limited Philippine immigration to only one hundred Filipinos per year. The only exceptions to this quota were for wives of US military personnel, active military, nurses, and others who qualified for two years of educational exchange. The 1965 Immigration Act radically changed this situation, eliminating national origin quotas and facilitating family reunification. But even then the waiting period was long, and there were many, like my uncle, who were never able to get a visa.

In his third year in medical school, my father saw other students take the foreign medical exam, which would allow top scorers to qualify for educational exchange. He became seduced by the prestige of foreign affairs. He wanted to have the same "bragging rights" as his colleagues headed to the States. So he took the exam in his fourth year, passed, and was accepted into an internship program at the University of Pittsburgh.

My mother could have also qualified for educational exchange, as the pharmaceutical industry in America was growing and in need of a new labor force. But her route to America was altogether different. She entered the United States on a tourist visa, as part of a trip sponsored by her mother, to allow her to "expand her horizons" before she got married. While in New York, she decided to stay to continue her studies. She wanted to pursue a PhD in industrial pharmacy to teach or to perhaps become an executive at one of the booming companies on the East Coast.

Thanks to American education in the Philippines, Filipinos like my parents provided a large English-speaking workforce who could easily assimilate while still taking positions that most Americans would rather not do—such as nursing jobs on the frontlines of COVID care, domestic labor, or doctor posts in neglected rural areas. This was an example of the infrastructure of colonial migration that pulled workers from the Philippines to the United States to address labor shortages, previously in agriculture and more recently in the American health care industry. This mighty pull factor seeded the influx of Filipino doctors and nurses to the United States that grew significantly after 1965.

American colonizers knew how to design a return on their twenty-million-dollar investment.

VII.

I have been pulled back to the main library by the force of inertia. I still don't know where to find food, so I consider finishing my research and calling it a day. But my neck is too constricted from bending toward books, and my sacrum is too sore from sitting. My gaze is lured instead to the bustling streets beyond campus, so I turn away from the library to walk to the corner of Katipunan Avenue and Shuster. I've studied the jeepney routes and know where to board the campus jeepney heading south.

"Never, never ride a jeepney," my mother once told me. "There is nowhere more dirty than the back of a jeepney."

"Jeepneys are dangerous," my father added. "When I was a med student, two men jumped me in a jeepney. One held a blackened bayonet to my throat, and the other one grabbed my wallet and my Hamilton watch. I could have been killed."

This assault is one of the reasons my father decided to go to the United States. What would have been the point of studying so hard to move ahead if it all ended in the back of a Jeep?

It's been seven years since my parents delivered these warnings. By now I am less afraid of what they fear than I am of the depression and anxiety I feel from avoiding all the would-be dangers in the world. I am willing to take my chances. I stand at the corner and peer into the distance. Two workers jostle a wheelbarrow overburdened with wet cement. Students lounge in the shade, their white teeth flashing at jokes and tsismis. I lean against a fence and shift my posture to feign like I belong.

Finally, the Katipunan jeepney approaches. With effort, I wave my hand toward the pavement. The jeepney comes rumbling to a stop. I step off the curb and approach its back door, or its lack of a door, the space where a door should be. Instead of a door, there is an opening.

VIII.

When my father entered the United States in 1962, he came through Seattle at the time of the Century 21 Exposition. As he flew over the city, he saw the newly constructed Space Needle and remembered feeling awestruck at how far it moved into the sky.

After clearing immigration, he went to Seattle Center to attend the exposition. There, he became overwhelmed by all the new buildings and clean sidewalks. And the food! At a stand selling waffles, my father marveled at a counter full of toppings: fresh raspberries, bananas, whipped cream, sliced almonds, butterscotch chips, strawberry syrup.

This was the America he had been educated to expect: so many choices, so much abundance in the promised land!

Then his face fell. Each topping cost twenty-five cents. He was counting his pennies with only a few bills of spending money to last him the whole trip out east. He had to eat his waffle bare.

Perhaps while he ate my father daydreamed of a waffle piled high with every topping on offer, like an edible monument to American excess. I imagine then and there, he vowed to himself that, in America, he would become a man who could afford it all.

IX.

I haven't checked to see if I have appropriate change for the jeepney, but it's too late now. I'm already on board and it's almost full. I take a deep breath to release anxiety and excitement. *You can do this. No one is going to pull out a bayonet. You are safe. You are coming back home.*

I see a sliver of red cushion between the jeaned legs of a teenage boy and a young woman wearing impeccably ironed green pants. I try to gracefully drop myself into the tiny space, and end up betraying my rookie status, stumbling as the jeepney lurches forward and plopping a butt cheek on the young woman's lap.

"Sorry," I mumble. She smiles, says, "Don't worry," and smooths out her pants. I ignore the throbbing pain in my tailbone, relax into the support of her left shoulder and the teenage boy's right arm, and smile. A man across from me smiles back.

This is my mother's worst fear. Close encounters of the human kind. Flesh on flesh, the rank humidity of human sweat, and the dirt of strange hands pressing palms with warm coins. Vulnerability, danger, disease. This is also my father's worst fear. Entrapment, ties that bind, obstruction of escape routes to higher ground.

X.

In America my parents found a land of cleanliness and space, and they also found each other at the wedding of a mutual friend from UP.

My father was in Chicago for a general surgery internship, and my mother happened to be there as well. She had stopped her tour to spend time with her godparents and to work under the table as a hospital phlebotomist.

My father said that he fell for my mother's beauty and how she played the piano "like a dream." My mother said she liked my father because "he was tall." My father was the romantic, and my mother was the practical one who associated height with advantages in life.

They married six months after they met, at St. Jean Baptiste Church in New York City. My mother's brother traveled to New York on a visa to give her away and ended up staying for good. That same night, my parents packed all their belongings in a rented Ford Galaxie 500 and drove to Niagara Falls for their honeymoon. It was Labor Day weekend 1964, and all the motels were booked, but they managed to find "a hut," as my dad called it, where they passed the weekend before driving to Pittsburgh.

In Pittsburgh my father began his neurosurgery residency and bought his first car: a used 1957 black Thunderbird. He was so proud of it, he washed that car and waxed it, and it looked good but only got five miles per gallon. It also stalled as often as it ran. His friends called it "the Pushmobile," but my father called it his "Sweet Blackbird." He put forty or fifty a month on it till he paid his first car off.

During their first year together, my mother kept house in what she described as their "dingy" apartment, which she put up with because she had no other choice. That's when she became a full partner in my father's plan for upward mobility. When the chief of neurosurgery at the University of Pittsburgh, a white man, openly questioned my father's capabilities as a future neurosurgeon, my mother clapped back. The chief believed my father's Philippine medical education was subpar. So my mother devoted her Philippine scientific education to reading and summarizing medical textbooks for my father. My father in turn studied these notes and applied them when on duty at the hospital. With the aid of my mother's

CliffsNotes, he eventually won over the chief, who then treated him as an impressive foreign medical graduate, an exception to the rule.

After three years in Pittsburgh, my father's mentor, Dr. Hubert Rosomoff, invited him to come to the Albert Einstein College of Medicine in the Bronx, where Rosomoff had been named chief of neurosurgery. It must have been a significant opportunity, since a white resident accused my father of receiving preferential treatment.

For a year after they moved to the Bronx, my parents enjoyed the big-city life. My father conducted research and had a light schedule, so whenever he could afford the tickets, he took my mother to concerts at Lincoln Center and Carnegie Hall. He won an award for his research on a surgical technique for relieving intractable pain in patients with spinal cord problems. The technique was a precursor to modern-day ablation. And then he started two years of trauma surgery, during which he worked almost round the clock. Meanwhile, my mother had a job at the Albert Einstein Microbiology Department studying *Streptococcus pyogenes*, the bacteria responsible for strep throat.

When he finished his residency in 1969, my father was supposed to go back to the Philippines, as was a requirement of his J-1 visa. My mother would have had to go with him. But Dr. Rosomoff petitioned for him to stay, citing that his services as a neurosurgeon were "needed badly." The Albert Einstein College of Medicine helped him become a permanent resident, then hired him as an instructor of neurosurgery.

And so my parents earned citizenship on the strength of their STEM skills and English fluency, and because my father filled a need in the rapidly expanding industry of American health. In the process, they became emblems of immigrant mobility and striver class excellence. They also became poster children of America's so-called benevolent colonization, witting examples of the US civilizing mission's success.

XI.

I've dug some coins from my pockets and passed them forward to pay for my ride. The Katipunan jeepney picks up speed, and the colors outside blend like an impressionist painting, throbbing with lovely brown life.

We arrive at the Katipunan Jeepney Terminal at Aurora Boulevard, where I will once again be faced with a choice of where to go. The man who smiled at me when I boarded has amazing stamina. Half an hour later, he is still smiling when he says, "Have a good day, ma'am" as we exit the jeepney. "Thank you, po," I say, using the formal address as I follow him out. I do a reverse L-shaped shuffle and step down from the bumper into the belly of the transit beast.

Under the Katipunan "flyover"—as overpasses are called in chronically traffic-clogged Manila—the jeepney terminal is its own metropolis of activity. Pedicabs, tricycles, commuters, and vendors fill the space between queues. I cover my mouth with a handkerchief to filter the briny smell of diesel, and now the other commuters and I are in a dance—forearms grazing triceps, fingers brushing hips, breath and sweat mingling to create an atmospheric glue binding the fragments between so many bodies. I absorb the sensuality like a newborn suckles milk.

I have completely forgotten my quest for lunch. My perception of bodily pain has been dulled. All I want is to be immersed in this beautiful brown crowd, which carries me along like a river.

XII.

After New York, my parents moved with my young brothers to the southwest suburbs of Chicago. There, we tucked ourselves into the heartland of America, as if we had grown there, sowed and rooted like cornstalks. There, my brothers and I lived as if we had never come from another land, as if my parents had never crossed the Earth's largest sea.

We moved into a red brick house off a dead-end road at the edge of a wetland preserve. We had no friends or family for miles. There were no grandparents in the spare bedroom, no relatives crashing till they got on their feet, no aunts and uncles coming to visit or cousins sleeping on the couch to get away from their parents for a while. Instead of an intergenerational household and an extended family base—both of which are common in Filipino families for reasons of both culture and class—we had a nuclear-family household free from extended family ties. I believe my father thought of this as progress. He had all but cut himself off from connection with his extended family and shaped domestic life in the model of white suburban America.

My parents' move to what their New York friends called "the hinterlands" of Chicagoland was sparked by the rising cost of malpractice insurance in New York State. In New York my parents had a large community of friends, and my mother had the time to bring my brothers to the park, to music lessons, and to Montessori. I believe they could have stayed there and lived a comfortable life. But for my father, comfortable wasn't enough. He wanted to continue on his path of American upward mobility, in a direct line toward the sky like the Space Needle he had seen upon entry. And so in Chicago, he paid lower malpractice insurance, invested more money, and joined forces with another UP medical graduate to dominate the field of neurosurgery in southwest Chicagoland. With the additional money that grew from his Chicago career, he shared, exerted, and advanced this success by hiring a housekeeper to cook and clean.

Lydia Servano was a family friend of my mother's mother, Lola Simeona. Lydia moved in with us in 1980, and over the next ten years she helped make my parents' American Dream come true, freeing up my mother and father to work long hours and earn even more. Lydia put hot food on the table for us and kept our big house clean. Meanwhile, Lydia sent much of her wages back home to her family in Iloilo. The money she sent helped relatives go through school, build houses, and come to America. But the cost of this mobility was great, especially for Lydia. In the ten years that she worked for our family, she only went home to see her own family twice. Like many of the four million Filipina domestic workers in the global diaspora, Lydia sacrificed time with her loved ones and also the chance to have her own family in order to take care of an employer's family—in this case, ours.

Growing up in Tondo, Manila, my father wasn't raised in a household with domestic help. The fact that he could now employ a worker was more evidence of his success, part of his ongoing trajectory of upward mobility. But he didn't stop there. Though a lower malpractice insurance rate should have meant he could work less than he did in New York, my father instead worked far more. At his peak, he reported to five different hospitals across Chicagoland. His extreme schedule led to a near heart attack that forced him to retire at fifty-five. It was almost as if he couldn't rest. Perhaps colonial mentality had taught my parents that you had to run at life like it was a zero-sum game. You were either a master or a subject, a winner or a loser; there was no in between.

XIII.

Under the columns of Katipunan Avenue, I dart between tributaries of moving people toward the jeepneys bound for Marikina. The lines are long but move steadily as each queued vehicle swallows throngs of people at a time. When it is my turn to board, the jeepney that approaches is a shining black diamond. The metal plate on top of the windshield is adorned with hot pink graffiti that reads "Loverboy."

I nearly squeal with delight; my wandering has been rewarded by a patók jeepney. "Patók," I had learned from my cousin, is slang for "a winner," a sure bet at the races. True to form, Loverboy has glossy enamel paint with a spotless chrome grill, four silver trumpets affixed to the hood, and airbrushed paintings of faces on its side, presumably the driver and his family. There's a zebra-striped boomerang affixed to the roof, and vinyl fringe dripping from its running boards.

I board through the back, buoyed upward by the surge of excitement in my chest. Inside, it smells like lemon air freshener. Within seconds, Loverboy is full to the brim. I am sardined between a group of girls my age and a group of boys who look no more than twelve. A couple teenagers in the back have their legs hanging out the door.

Bump, thump, whirr, and *clank,* Loverboy revs to life. All of us begin to bob our heads as Blackstreet blares from the stereo system while coins clink in rhythm from hand to hand. *Clink, clank, clink, clank,* "Bayad po," "Bayad daw," "Sukli." Cold metal, warm sweat, thighs pressing thighs, shoulders otso-otso to the beat as palms move like a bucket brigade in a collective ritual of paying our fare.

XIV.

In our house on the hill, one of the most prominent photos on display was of Ronald Reagan. The photo was framed and signed by the Gipper himself. It was a testament to my parents' conservative economic values and their trickle-up accrual of wealth.

My mother had closets full of fine clothes she covered in plastic to preserve. She pinned handwritten notes to the plastic sheaths, detailing the date she last wore the outfit and to what occasion. This system ensured that she wouldn't wear the same clothes in front of the same people twice.

My father had a garage full of what he thought to be the quintessential American car: a lipstick-pink Cadillac, embedded with sparkles like I remember seeing at the Wet N Wild displays in Osco Drug. Inside, the Caddy's leather was pure white, not beige or taupe, but white like milk and soft like Jet-Puffed Marshmallows. The car was so luxurious I sometimes crawled into the back while it was parked, just to take a nap. And it was so big that the garage wouldn't close unless we parked it within an inch of the door.

Over time, my father had become a man who could afford it all. But as my parents got more successful, we also grew more isolated. The more we had, the more they believed someone would come to take it all away. And so they erected a six-foot-tall iron fence around the property and kept the gate tightly closed. They rarely allowed my friends to visit—friends were not to be trusted. They kept extended family and even members of their own social circle at a safe distance.

For our family, isolation was the consequence of upward mobility. Separation from others was the final destination of colonial migration.

Much later, I learned that my parents moved an invisible hand to support family members in the Philippines. Judging from the fact that we had little to no contact with these family members growing up, I believe my parents treated these as business transactions with few emotional ties.

They didn't take time to do what many Filipino American families do: pack balikbayan boxes full of American-made canned corned beef and peanut M&M's and socks and vitamins to send home. But they wired money to support my uncle and to put my cousin through nursing school. They put another cousin—the daughter of my uncle's second "common-law" family—through college, with no judgment against her for being an "illegitimate" child. They sent money home to other cousins to help them buy a condominium in Manila. Through my own trips to the Philippines, I have become close to this group of cousins, who still live in that same condominium today.

I.

Loverboy has turned onto Andres Bonifacio Avenue and is now weaving rather merrily through the thicket of traffic. We cross the Marikina River, the largest tributary of the Pasig River which is the main waterway of Metro Manila. I catch a glimpse of the Marikina River Park with its placid stretch of brown water flanked by wide stairs on one side and a greenway and walking path on the other. The park looks pleasant, but the smell from the river makes my nose wrinkle. I have read about this park, a sort of oasis in the concrete jungle, a deliberately grown green space amidst a city sprung like weeds from the rubble of World War II.

Loverboy is now less than half-full as many of its passengers disembarked at the river park stop. With a whole foot of space between the next rider and me, I learn something that now seems like a law of physics. I realize that when jeepneys are packed, as they are meant to be, you can ride in relative comfort, wedged among and cushioned by your fellow riders. This cushioning even felt like a tourniquet for my pain. But the thing about jeepneys is that they don't work unless there is a village of people in back. If they are largely empty, as Loverboy is now, your ride turns to hell. Alone, you have no purchase. You fly from seat to floor, from the front end of the bench to the back. You try to hold on, but the sides of open windows don't give much grip, and you realize as you try to brace yourself by pressing your knees together and pushing your feet to the floor that those inner thigh muscles you never use are too weak to save you now.

I hit Loverboy's ceiling twice with my palm–rap, rap–and say, "Para po" to the driver. Loverboy screeches to a stop. The momentum throws me forward against the wall and then backward toward the open door.

Out in the street, I am once again enveloped by the warmth of a sensual crowd. I watch Loverboy hurtle away and silently thank it for the ride. Across the road, I buy a cone of peanuts for five pesos. This was one of my father's favorite things to eat growing up in Tondo, and while I savor the nutty smell and the fresh-roasted heat on my tongue, I imagine the nostalgia he would feel if he were here, back home, eating with me. I picture him consuming the peanuts with pride, then launching into a story about how he is one of the survivors who made it out of Tondo.

I no longer feel the need to search for lunch. The peanuts are enough. The unsheltered light of this afternoon sun has pulled me to somewhere beautiful. Immersion in collective life has left me sure of where to go.

Reverb echoes from the subwoofer bass as I board a new jeepney on the other side of Andres Bonifacio Avenue, bound for Katipunan. Riding comfortably tucked in among strangers, I circle back to the university campus, back toward an ancestral history of colonization and revolution—the direction from which I came.

II
Neural Pruning

- The biological process of reducing neural connections to those most used, to increase the efficiency of synaptic signals. Otherwise known as "use it or lose it" and "neurons that fire together, wire together."
- Pulling back the lens, to trace how trauma and resilience can be transmitted across generations.

We threw them to the ground and pinned their arms and
legs, we kept their mouths open with bamboo sticks,
we poured water in till they almost strangled,
as the land often strangles with the monsoon.

And then we jumped on them to let the water go.

—ERIC GAMALINDA

War-Fire

But the war goes on.

—FRANTZ FANON

F2: *apo*

When I was thirteen, my family boarded a 747 from O'Hare to Manila to visit my grandmother, my lola Simeona. It was 1989, only three years after the People Power Revolution deposed Ferdinand Marcos and ended martial law. But my lola wanted to tell stories of a different war, the Second World War, in which the Pacific theater lay in the long shadow of the crown and in the crosshairs of the eagle and red sun.

On that trip, Lola took me out to her lanai to tell me a story. I sat next to her and held her hand, and I remember her skin was like onionskin, paper-thin, with blue veins running like rivers above delicately carved fingers of knuckle and bone. As she shared her kwentos, I gripped her hand tighter, then released it quickly with the fear that it might break. I did not yet understand her resilience, her fire-borne strength.

F0: *lola*

When she was eighteen, Lola Simeona sold food at the sabong. As a vendor, she was one of the only women permitted inside the arena. My grandfather Lolo Juan came to the cockfight with his father, who saw Lola and became immediately entranced by her. She moved gracefully as she worked, spooning rice and ulam and exchanging banana leaf–wrapped packets of suman for coins. To my great-grandfather's eyes, she was enterprising and also lovely, so much so that he dragged his son over to meet her, immediately convinced she was the one for him.

Soon after, my grandfather went to my grandmother's house to court her. He found her two younger brothers at home alone, tied to a pole with rope. My grandmother was out gathering firewood. She had tied the boys up to keep them out of trouble while she worked. The next day, my grandfather returned, this time with a gift: a large stack of firewood he had chopped himself.

My lola and lolo married after a brief but romantic courtship. They had three children: two boys and my mother, the bunso, who was born in 1939. And then, World War II rained down on the Pacific. On December 8, 1941, just ten hours after the bombing of Pearl Harbor, the Japanese Air Force bombed eight targets throughout the Philippine archipelago. American forces in the Philippines were caught unprepared.

Lolo Juan left immediately to enlist. Meanwhile, Lola stayed home to protect the family.

I shifted in my chair as I listened to my grandmother, nervous about her and my lolo's safety. I took a deep breath. My nostrils filled with the scent of orchids mixed with diesel exhaust from the street. Lola sat straight in the wicker chair beside me and continued her story in a steady but quiet voice.

In late 1942 or early 1943, Lolo Juan and two other officers staged a tactical operation to allow their commanding general to escape enemy forces. The operation was successful, but the Japanese Army captured my grandfather, and he became a prisoner of war.

Lola's voice turned bitter. I shivered in the humid heat and looked out at the yard beyond the lanai. Before the war, Lolo Juan had seeded mango trees that now twisted two stories into the sky. Lola had grafted orchids onto their branches. The orchids bobbed, coquettish, splashes of white and purple and tendriled green upon the gnarled bark of the trees. Lola fixed her gaze on a faraway place. Her voice reduced to a whisper, and I had to lean closer to hear her words:

> When the Japanese captured your grandfather, they took him to Intramuros and tortured him. They took him onto the hill where he could see the Pasig River, maybe so he could

observe how close he was to his country, but so he would know that he was not free.

Lola visited Lolo Juan in prison. She did not explain to me how this came about. Looking back, I wonder if Japanese officers took her there, either as a kindness or a form of psychological warfare. Maybe my grandmother demanded it. Or perhaps it was a transactional exchange for her body and her home.

> They filled his stomach with water like a balloon. Then they would take turns jumping on him again and again and again until he threw up blood. They wanted him to tell them who were his fellow soldiers and where they were in the mountains. But he never told them. He never told them where they were hiding or gave them any names.

I winced at her description of my grandfather's torture. Unconsciously, my hand reached up to my own neck, to a place near my throat that has always felt silenced, near drowned. I kept my palm there to brace myself, and focused back on Lola. I noticed the folds of skin on her face, how they hung in a way that made her appear perpetually sad. Yet I could also see pride in her eyes, a flame that smoldered when she concluded:

> The war ended, but your grandfather did not come home. His body was never found.

I was stunned. My heart ached for Lola and for the grandfather I would never know. When a well of tears overflowed from her eyes, I squeezed her hand tight once more.

F1: *anak*

My mother never spoke to me about my grandfather. I was an adult before I even learned his name was Juan. When I asked her questions about him, she could only remember that he was a strict disciplinarian who always told her to chew with her mouth closed. She also recalled one story

Lola had told her: Lolo Juan expected her brothers to be home by 6 p.m. every day, when the church bells rang. One afternoon, her brothers were having such a good time diving and swimming in the river that they didn't hear the tolling of the bells. Lolo Juan had to fetch them, and once he'd dragged them home, he whipped them both, saving more strikes for the elder, because according to Lolo Juan, he should have known better.

Beyond this, my mother doesn't remember him much at all since he left for the war when she was around three. With my mother essentially fatherless, it was as if the memory of him skipped a generation. I believe Lola told me her war stories because she wanted to make sure that another part of Lolo Juan lived on through me.

F2: *apo*

One evening in 2020, my husband was in the bathroom when he let out an anguished cry.

"Juan! What happened?" Panicked, I burst through the closed door to find him doubled over in pain.

"I hit my finger on the counter, and it hurts like hell," he said. A few weeks before, he had fallen and fractured his right pointer finger, which was now wrapped in a soft splint.

"Ouch!" I winced. "I'm gonna go get you some ice."

I gave him a hug, walked out of the bathroom, and proceeded to collapse on the hallway floor.

Now it was Juan's turn to rush over to me. "Are you okay?" he asked, alarmed. When I didn't respond, he asked again and again, and again and again I could not answer.

I was facedown on the floor, my torso trembling, my head shaking uncontrollably. Suddenly, words began to pour from my mouth like water from a broken dam: "I couldn't do it, I couldn't do it, I couldn't do it!"

I repeated this over and over while pounding my fists on the floorboards. The repetitive phrase, the tears, the sudden despair not only overwhelmed me, they *were* me; they were my mind, my body, my memories—but not my own.

Juan used his good hand to help me to bed. He wrapped the covers around me while I rocked like a baby and chattered:

"I couldn't do it. I couldn't save him. I couldn't save him, I couldn't save him. I couldn't save him."

"I couldn't save him!" This last one came out like a wail.

Juan stayed with me until I fell asleep from exhaustion. For days afterward, I was fatigued. My body ached so much, I thought I could hear the pain like sound waves reverberating from my flesh. I felt a commanding urge to hide, to armor myself, to grieve.

I had once experienced an implicit flashback to infancy, an embodied memory of what could have been trauma from my own birth. But this time, my body knew that the episode I reenacted was not one that I had ever experienced myself. While it was happening, all I could think of was my lola Simeona and my lolo Juan, and how they had found each other at a cockfight and lost each other in a war.

Could this have been a transgenerational flashback? I had heard of past-life regressions, but not of implicit ancestral flashbacks. It sounded crazy in my mind, but I began to believe that I had experienced an archival memory of what my lola might have felt when she saw Lolo Juan imprisoned, or perhaps it was an embodied residue of what she felt after the war, when my grandfather's body was never found.

F0: *lola*

In the 1970 movie *Sunflower*, Giovanna, a young Italian bride, goes to the Soviet Union to find her husband, who goes missing in action during World War II. Giovanna visits the sunflower fields, where there is one flower for each fallen Italian soldier and where Germans have forced Italians to dig their own mass graves. Eventually, she finds her husband alive but with amnesia. The war made him forget his former life as well as the horrors of combat, at least in his thinking mind.

My cousin tells me that Lola used to compare herself to Giovanna. After the war, the Bocaue town council wanted to honor my grandfather by naming his childhood home the "Barrio of Juan San Pedro." But my grandmother refused. My cousin thinks she still couldn't accept that he was gone. She wanted to believe that like Giovanna, she would one day find her lost husband, or that he would one day find his way home.

When my grandmother told me the story of losing Lolo, my thirteen-year-old self also began to hope that somehow he might still return.

F1: *anak*

When my father died in 2017, my mother spent weeks at home alone, crying. My brothers and I encouraged her to go see friends, to go to church, to call the grief support hotline left by the hospice nurse. But my mother refused, saying through tears that she didn't want to see or talk to anyone because all she would do is cry.

Somewhere along the way, my mother had learned that fluvial emotions should not be shared. She had passed a similar message on to me as a child. Instead, she found an outlet in sad Hallmark movies which would allow her to cry without the demands of human interaction, in the privacy of her own home.

For the past five years since my father died, my brothers and I have been trying to help my mother move to Seattle to be near us. Yet she continues to return to Las Vegas, to the home she shared with my father. She is attached to his memory and to the material goods they earned. But beyond this emotional link to my father, I believe that my mother practices nonattachment to other human beings—perhaps to balance out her worry over being robbed of things. I wonder if this is a primal defense against the terrible human losses of war. One need not grieve that which has never been beloved.

Recently, I told her that my son had been asking when she would come back to Seattle to visit. She replied that she probably wouldn't be back anytime soon. When I told her that would make him sad, she said, "Oh, it's better if I don't spend too much time with him, that way he won't miss me when I'm gone."

F2: *apo*

Unresolved grief is a legacy of colonization and war. Over time, it can wear at the soul like uncontained floodwaters can erode earth, long after the end of the storm.

After the incident where I felt I was reliving my grandmother's grief,

I came across the work of Lakota social worker Maria Yellow Horse Brave Heart. Her research extended my journey to tracing the roots of trauma beyond my own life.

Brave Heart has shown that the clinical definition of PTSD has limits when it comes to people with ancestral trauma from enslavement, colonization, and war. PTSD, and even complex PTSD, deal with only lived experiences. But for Native communities, Brave Heart saw that a new framework was necessary to encompass the disruption of threat response across generations. She developed the framework of historical trauma, which carries with it the related experience of unresolved generational grief.

As she describes it, for Native people, a transgenerational perspective is necessary because of the cultural lifeway of ancestral connectedness and because their communities have been hit by so many losses for so long. "We didn't have time to heal from one loss before another occurred," she says.

When one generation is simply preoccupied with survival, the trauma response continues, and often worsens, in future generations, until the trauma can begin to be processed. Only when we face this historical trauma response—which can range from grief and illness to depression and addiction—can healing begin both for the individual and for the generations that come before and beyond.

F0: *lolo*

As I write, I look at a sepia photograph of my grandfather and recognize my set jaw and knowing smirk in him. Lolo Juan was a handsome, dark-skinned man who, before meeting my grandmother, was considered the most eligible bachelor in town. He came from a family of educators and was a school superintendent in the American colonial education system before he enlisted. After he and Lola married, Lolo Juan built a family home outside the town of Bocaue in the barrio, next to the street. He crafted the house from local wood, and when finished, it stood two stories tall with Spanish-style balconies and capiz-shell sliding windows. The house was so beautiful that after the Manila area fell to the Japanese, the Imperial Army thought it the perfect headquarters from which to base their regional mission: to neutralize guerrilla units—like Lolo Juan's—resisting Japanese occupation in Central Luzon.

F1: *anak*

It is the summer of 1942, months after my grandfather left to join the guerrillas, but at least one year before he becomes a prisoner of war. American troops in the Philippines officially surrender the archipelago to Japan. American commanders retreat to Australia, leaving Filipinos like Lolo to fight the Imperial Army and Filipinas like Lola to salvage what they can of domestic life.

My mother is three years old, and there are strangers banging on the front door. Lola tells her to stay behind her, shielding my mother's small body with her own as she walks slowly to address the moment she knew would come.

When Lola opens the door, she finds a group of insistent Japanese officers. My mother stands between them and the bookcase, where the family money is hidden in a brown paper bag. My mother is only a toddler, but she already knows, as if by instinct, how to protect what is needed to survive.

F2: *apo*

I imagine my mother as a three-year-old, a skinny toddler with legs planted firmly on the ground, mustering all her strength to stop grown men from stealing the family money. Neither my mother nor my grandmother ever told me what happened after Japanese soldiers came to their door. Whether it's because they can't remember or choose not to, I don't know. I am left to wonder about the emotions of that day and how their impacts may have reverberated for decades to come.

The egg that became me was already in my mother's body during the war, when she protected the family money, when her father left to fight and never came home. It was already there, in fact, when she was born. This egg that became me, like all egg and sperm, was sensitive and may have changed in response to stress and trauma in her environment.

The period between three and sixteen years of age is a sensitive time for neural pruning and may also be a key period for epigenetic changes—or changes in gene expression—in germ cells. How might my mother's body have adapted at such a sensitive period in her life? How might my mother's eggs have been altered from the stress and trauma of war? What might I have inherited from these adaptations?

F0: *lola*

Long after World War II ended, my lola Simeona told my cousin that she would have used a bolo to kill any Japanese soldier who dared enter her home. In retrospect, my cousin and I don't know what to make of this. We don't know whether this was wishful thinking or a description of what Lola actually tried to do when Japanese officers came to her door.

When Lola told me stories about the war, she was more interested in conveying details about my grandfather than she was about disclosing details of what happened in her enemy-occupied home. Her movements after that summer of 1942 remain a mystery to me. What did Lola Simeona do when the Japanese soldiers entered the house? Did she try to kill them? Did she resist and meet with retaliation? Or did she acquiesce for the sake of her children but sharpen her bolo while the officers slept at night?

F1: *anak*

The Japanese officers have entered the house, and they approach my mother where she continues to guard the money. She successfully diverts their attention. Or she does not, and they raid the house and take the cash. The Japanese officers must know that my grandfather is a guerrilla fighter and that there are no adult men in the house. There is only Lola, my three-year-old mother, and my six- and seven-year-old uncles. The officers take advantage of this situation, settle in, and occupy my family's home for the duration of the war. Or perhaps they only stay for a short time.

The battleground of war extends to the battleground of memory. My mother only knows that she used her own little body to guard their money. And she recalls a moment when she broke her arm while playing in the rice fields. A Japanese soldier set it for her so that it would heal.

Overall, she says that if her memory serves her right, the Japanese soldiers treated her well.

F2: *apo*

I am ten years old and know nothing about war. I have not yet heard my grandmother's stories and can glean no details from my parents' silence. And yet I live daily life preparing for battle.

It starts with the fight to get out of bed. I fend off the heavy monster who sits on my chest at night in order to lift myself up and begin the day. I throw back my bed covers, swing my legs down to the floor, and tiptoe into the morning. I am thankful that I have lived to see another day.

I spend the rest of my waking hours on high alert. I brace my body against the possibility of ridicule, scan the schoolyard for possible threat. I hide from bullies. But I myself am one of them. I flush a classmate's sweater down the toilet, stick a maxi-pad stained with red marker on her locker, make fun of her foreign-sounding name. It feels better to attack than be attacked. Thankfully, I outgrow this when my sensitivity tells me it's wrong. But I lie in wait for other opportunities to turn my fear into aggression.

From my bedroom window, my eyes dart across the yard, seeking any trace of intruders. I ready myself for another night of savage stuffed animals, creaks from the attic, and violent recurring dreams. I know when I close my eyes I will once again be chased by unknown attackers with bolos for hands, and once again I will be caught, and when I try to fight back, my arms will be soft like rubber.

Before bed, I choose one of the weapons from the wall in my brothers' room—a stainless steel throwing knife, a Chinese star, or a balisong blade—and I tuck it away between my ruffled bed and the flower-papered wall.

F0: *lola*

What exactly happened in my grandmother's enemy-occupied home? To think of it makes me shudder. Not every Japanese soldier raped women, but some of them did. Lola Simeona birthed a third son in 1943, two years after the start of the war. This uncle was not beloved by my grandmother and died young of liver failure from drinking too much. Even my mother rarely mentions him, and when she does, she calls him the "black sheep" of the family. When she talks about her brothers, she says that she had two, not three.

The details of what happened in that house in those years are lost to history and can only be invented through imagination. But the context of Japanese brutality during the war is well-documented. More than 1.1 million Filipinos were killed between 1941 and 1945. Daily life under Japanese occupation was characterized by brutality and war crimes. The Imperial Army demanded that all Filipinos bow to them or be slapped. Japanese soldiers impaled babies on bayonets, battalions burned whole villages and skinned civilians alive, and thousands of young women were raped, with many held prisoner and forced into sex slavery by Japanese officers.

Some of the eight guerrilla units that had formed in Bulacan came together in direct response to Japanese atrocities in the area, including an increased incidence of seizing harvests from farmers and sexually assaulting women. My grandmother doesn't speak of any of this. This was not part of the kwentos she shared on her lanai. And so I give her the benefit of insinuation over declaration. Through her, I have inherited drops of stories, but the cup in which I gather them remains half-full.

F1: *anak*

When my mother says she remembers that the Japanese soldiers treated her well, I think of the movie *Life Is Beautiful*. In the film, a Jewish father and his young son are sent to a concentration camp in Nazi-occupied Italy during World War II. The father creates an elaborate game for his child to shield him from the horrors of the camp and to help him survive with a sense of joy. The boy lives and is reunited with his mother.

I suspect Lola did a similar thing for my mother when the Japanese army took over their house. Perhaps Lola took my mother outside to the rice fields to play hide-and-seek. Perhaps she turned the radio on high so my mother could listen to Perry Como sing "Till the End of Time," and maybe the music drowned out the sounds of drunken officers or plans of attack or whatever else may have been happening inside that home.

While details may be lost to history, emotional archives live on within the bodies of the survived. Even though my mother remembers kindness from Japanese soldiers, and my grandmother likely did everything she could to shelter her only daughter from the horrors of the war, my mother's bodily state has always seemed to me to be one of hypervigilant defense.

She lives in constant worry that at any moment, something may be stolen from her. On a recent family trip back to the Philippines, my mother couldn't find a necklace she had left on her night table. I began to look for it, but she said it was no use; she was sure it had been stolen. She sat down to cry. Minutes later, I found the necklace inside her purse, which was on the floor next to the night table. I told her it must have fallen. She looked at the gold chain like she couldn't believe her eyes, then wiped her tears away with relief.

In 2022, she fell and broke her pelvis while doing laundry in her house in Las Vegas. The next day in the hospital, she worried less about her injury than about her conviction that a paramedic had gone back to take valuables from her home. She had my brother change the locks immediately. To keep neighbors from looking in, she continues to close her house blinds during the day, a habit I remember well from childhood.

Somatic therapist Resmaa Menakem writes of "trauma ghosting," which he calls a common and overlooked response to trauma. He describes it as "the body's recurrent or pervasive sense that danger is just around the corner."

My mother may be haunted by trauma ghosting. How much was taken from her during the war, that she must constantly defend against loss in old age? Did her father's lost body and my grandmother's grieving of him cause epigenetic changes, which in turn engendered reactive behavior as a means to survive? One of the very definitions of PTSD involves continued defensive reactions to something that has passed, even when those reactions don't fit current conditions at all.

Lola Simeona wanted my mother to have a pampered existence, and she must have succeeded since my mother describes her own life as "soft and easy." I'll never know the details of what Lola might have done to shelter her, but I see that even as my mother reacts defensively and holds herself within a protective cocoon, she also retains a childlike innocence toward many aspects of the world, even now.

F2: *apo*

If the son in *Life Is Beautiful* were a real-life person, he may have become a memorial candle. Psychotherapist Dina Wardi developed the concept of memorial candles to describe how the children of some Holocaust survivors carry on their parents' unresolved traumas. Neuroscientist Rachel Yehuda made waves by being one of the first to show the science behind this concept.

When Yehuda opened a clinic for Holocaust survivors at Mount Sinai Medical Center in New York, her team was surprised to find that adult children of survivors came to seek help more than the survivors themselves. In 2015, Yehuda's team conducted a study of these adult children and found that they had altered levels of stress hormones like cortisol. These levels were similar to those found in previous studies with Vietnam veterans. The adult children of survivors were also more likely to have PTSD than the children of Jewish parents who had not experienced the Holocaust. In other words, the children of survivors had a disrupted stress response similar to those of their parents and to veterans who had directly experienced war, even though the children had never lived through war themselves.

In 2020, Linda Bierer of Mount Sinai affirmed Yehuda's results through a larger study and also found that epigenetic changes to a gene associated with cortisol production correlated with highly vigilant behaviors. These epigenetic changes and behaviors were especially present in the offspring of mothers exposed to the Holocaust as children.

This research was some of the first to suggest the intergenerational transmission of trauma in humans and to posit a scientific mechanism behind why, in the bodies of some, the war does indeed go on.

F0: *lola*

Lola Simeona used to tie her sons up to trees and beat them. So my uncle told me, and my cousin as well. She left this out of her storytelling about the war. I try to picture her among the mango trees and the orchids of her yard. I imagine her tying them up and beating them amidst tropical abundance. Was this something she had learned from her own family, since as a young adult, she also tied up her brothers when she went to gather firewood? Or was it something she learned from Lolo Juan, who whipped his sons when they failed to come home? Or was it a traumatic retention from war and colonization, not the result of culture, but the consequence of generations of unprocessed traumatic events?

Whipping children was not unusual in my grandmother and grandfather's time. And yet I wonder if this may be one way that Lola exhibited her own post-traumatic stress from the war. When I picture her beating my uncles, I think of how she sounded when she talked to me about the Japanese officers who tortured my grandfather. When she told me how much she hated them, her voice was a rusted machete—bitter and dull but still deadly. Perhaps, along with her stories of Lolo Juan, she passed this machete on to me.

F1: *anak*

My mother received the white-glove treatment while my uncles got beatings. She was vigilantly protected by both Lola and my uncles, who knew she was the favorite but never seemed to hold it against her.

At my childhood home in Chicagoland, when someone uprooted my mother's prized peonies and left them for dead, her reaction was to retreat into defense—erect the iron gate—and to peacefully pray. As far as I could tell, she never tended toward physical aggression when upset or threatened.

And yet my mother was no pushover. When it came to what she thought was right, she would raise her voice, while my father was more wont to let microaggressions and social mishaps pass. When I was a teenager, my mother and I went to a hair salon where they also did nails, and my mom was indignant when she smelled strong fumes and saw a pregnant woman working in the poorly ventilated room. She asked for the owner and proceeded to lecture her about the dangers of acetone fumes for pregnant women and babies. She said it would be easy to open more windows and install more fans, and that it would be good for both the workers and the customers. If the owner failed to do so, my mom said, she would report her.

Back then I was embarrassed by my mom making a scene, but I feel proud of her now. Perhaps this is what she inherited from Lola Simeona and Lolo Juan: a militant righteousness expressed as a verbal crusade more so than a penchant for whipping. Perhaps I learned from my grandparents where I came from, and from my mother the vision of where I could one day be.

F2: *apo*

I don't know what it's like to be a veteran of war. I have not been on battlegrounds, stepped over bodies, used them as shields, aimed at flesh to tear it apart. And yet war lives within me.

Now when I envision my grandmother beating her sons after tying them to trees, I see the source of violence I sometimes feel within myself. I have mostly turned this aggression inward through negative self-talk and punishing internal diatribes. I believe this hostility has contributed to my chronic pain. These days, as a parent myself, and as the parent of a highly physical, strong-willed child, I find myself wrestling with anger whose heat can move from flicker to flamethrower in a matter of seconds. I work to channel it in constructive ways, so that I don't take it out on my child and so that I don't model physical aggression as an acceptable way of expressing emotions.

And yet . . . I have fight dreams. Flight dreams. I have thoughts while chopping eggplant on a sunny day. I think about how easily this knife could cut flesh. I scan the slice of street outside my window and mentally dare someone, anyone, to break into our home. At night when I walk, I stride with my head high, I puff my chest and gaze hard to challenge the few who walk around me to fuck with me so I can fuck with them. *Fight me. I dare you. Fuck with me. I dare you.* It would feel so good to let this caged rage free.

F0: *lola*

Lola remarried several years after the war. She had many suitors, including a "friend" who my mother thinks may have been her lesbian lover; an army general who wanted to take her to Washington, DC; and a priest who, in full regalia, said he would renounce his holy vows for her. But my grandmother would have none of them. The only one good enough for her to marry was a man named Conrado, a comrade-in-arms of my grandfather's. Conrado had fought side by side with my lolo Juan during the war. Perhaps marrying him was another way that Lola tried to keep my grandfather alive.

Though Lola trusted Conrado as a spouse, she trusted no grown man around my mother. So when my mother was ten or eleven, Lola sent her away to live at a Catholic boarding school run by Belgian nuns.

F1: *anak*

The way my mom tells it, she was glad to go. Lola had told her stories about how difficult it was to keep safe as a girl in the barrio. After Lola's mother died when Lola was only fourteen, village elders counseled her to avoid all men, even her own father. The elders gave Lola payo, or advice about how to avoid rape and incest, and told her the devil was at work possessing men to rape even old women they met at the well. This was my mother's own generational inheritance through story.

At St. Theresa's in Manila, my mother enjoyed school and the nun's rigorous structure. She also enjoyed being sheltered from male advances and the male gaze. Boys had been chasing her home since she was in the third grade. Being cloistered meant that she avoided this unwanted attention as well as the conflicts—both verbal and physical, I can only assume—that ensued between her brothers and their new stepfather.

As much as my mother is unafraid to pick fights with bosses who endanger their workers, she is generally avoidant of many things she considers to be threats. In addition to closing the blinds during the day, she applies one-way reflective paper to her windows so that when she does open the blinds, she can see out, but no one can see in. She bleaches and sterilizes all surfaces to get rid of germs. She rarely answers her phone. I wouldn't call it paranoia so much as hypervigilance and perhaps a hypersensitivity to potential harm. These behaviors continue in times of peace.

Coleen Murphy is a Princeton geneticist who has shown how behavioral avoidance can be passed on for several generations, at least in worms. Working with a type of worm called *C. elegans*, Murphy's team determined that *C. elegans* who get sick from eating a bacteria they normally like to eat will pass on a "memory" of this experience that leads their offspring to avoid the bacteria as well.

C. elegans do not teach their offspring to avoid toxic bacteria through lectures or stories, as far as we know. The behavior must be biologically transmitted. Murphy and her team found that the mechanism for this transmission was a small RNA molecule from the bacteria that is passed into the germline of the worm. This RNA then signals a change in the expression of a gene in a sensory neuron that leads the next generation of *C. elegans* to avoid the pathogenic bacteria, and so on, and so on.

Murphy and her team found that this behavioral protection from threat was inherited from F0, the generation that was exposed to the toxic bacteria in the first place, all the way through F4, four generations later.

F2: *apo*

I once had a tai chi teacher who stepped close to me and used my hair as an example of one step in a form. Without asking, he chopped his hand through my long hair. It was not an aggressive movement, it was slow and sly, and according to my body, therein lay its danger. I also noticed that he had never stepped this close to any of the male-bodied students in the class. I switched to a different teacher, and later found out he had several complaints of sexual harassment and even assault made about him.

When I was twenty, I went to Malaysia to work as a travel writer. While ambling down a deserted road, I saw a monk walking toward me on the opposite side of the street. There are a lot of monks walking the streets of Malaysia, but for some reason, this one set off my visceral alarms. I gave him an extra-wide berth, and as I passed him, he flung up his robes and began thrusting his naked penis into midair. I was carrying a pocket knife, and if I had been any closer, I might have pulled a Lorena Bobbitt as a pure reflex of self-defense.

This sense of neuroception, which psychiatrist Stephen Porges describes as the nervous system's ability to assess risk in our environments, may actually be heightened through physiological changes in our neurons. Studies show that animals can generate new sensory neurons when exposed to threat, which may help them detect and respond to danger more quickly in the future. Emerging research suggests these adaptations can be epigenetically passed on to offspring, a biological mechanism that can help children adapt to threats that parents experienced in their lifetimes. Applied to humans, this could include the threat of invaders, a subtly invasive tai chi teacher, or a sexually predative monk.

Are we actually X-Men? Maybe some of us descended from survivors of rape, genocide, and imperial war have mutated, so to speak, at least at an epigenetic level. Our ancestors may have given us the gift of lessons learned from their environments. Perhaps our vigilance is an heirloom, an inherited superpower that helps us move away from danger and toward what will keep us safe, or even what will help us thrive.

F3 Through F7: *Generations Beyond*

When Lola Simeona passed away from heart disease in 1998, she was the last living family member of her generation. Now my mother is the last of her generation, on both my mother's and my father's side. Those of us who are their descendants carry on their histories. Our bodies are archives of not just our own lived experiences but of our ancestors' as well.

At thirteen I believed it possible that my grandfather could still be found, even forty-four years after the war. Now I know that my grandfather's lost body can be found in me and that his thwarted fight for freedom seeks resolution through my work. I have often felt that I carry the current of my grandmother's survival and grief in my nerves. Part of my nonbinary gender identity comes from feeling both Lola Simeona and Lolo Juan's spirits intertwined in me. I now feel that I also carry my mother's sixth sense of threat detection, as well as my ancestors' transgenerational commitment to justice. Though I have struggled to temper the flame of so much fire, I consider all of these inheritances a blessing.

My child is also a blessing. Ever since he was a toddler, he liked to shoot things for fun, like many kids do. Now he is eight and obsessed with the military and often thinks, as I do, about strategies of war. He also seems to be chronically hypervigilant, easily agitated, and anxious, which I am guessing he got from me.

Here, again, neuroscience can provide hope. Murphy and her team found that in *C. elegans* who learn to avoid toxic bacteria, the behavior is reversed by the fifth generation. Our genetic expression, also known as our phenotype and behavior, changes in dynamic relationship to our environments. Survival adaptations that can become harmful in new environments can eventually be reversed. For humans, this means that the worst impacts of inherited trauma are fluid rather than set in stone. Given the right conditions, from the ashes of war we can rewire a new neural circuit for peace.

381 Years

Colonization: *the scramble to cut and consume bodies, or dividing the map into slices of bitter imperial pie.*

Balagtasan: *the clapbacks that re-member our agency, or a head-to-toe anatomy of history that leads our bodies somewhere whole.*

CROWN

This is where we begin: in the
name of the Father, the Son,
and the Holy Spirit.

We learned to cross ourselves when
consuming the Body of Christ. Now
the motion is like a reflex, almost
involuntary.

What became the Philippines is a
body of more than seven thousand
islands crossed by the Spanish empire.

We carried the cross as a symbol of
our sins—sins so great, Christ gave
his body to wash them away with
blood of nail wounds and crown
of thorns.

In 1521, conquistador Ferdinand
Magellan reached the archipelago
and planted the first Spanish cross
in Asia in Cebuano soil.

I thought the lesson learned was that
we would all come to a bloody end.

Among Magellan's sins: he wanted
too much. Black pepper, nutmeg,
and gold. Then: three goats, three
pigs, and three loads of rice. This was
the beginning of his end.

Some of us accepted the provisions
of the host and renounced Allah,
Bathala, and the Diwata.

When native tribes agreed to
give two goats, two pigs, two
loads of rice but not three,
Magellan declared war.

Others of us went to battle.

The great Datu Lapu-Lapu's warriors felled the conquistador with a
poisoned arrow. In the end, Magellan died for his need for the Trinity.

Were we uplifted or were we felled
by the Father, the Son, and the Holy
Ghost? I have long felt my body
haunted, its contents invaded, my
veins full with other people's blood.

After Magellan, other conquerors were sent by the Crown, but it
would take three more tries before
they reached the archipelago again.

Nature gods once inhabited our archipelago. They enchanted rivers and trees, moistened mountains with sky water.

In 1565, Miguel López de Legazpi followed in Magellan's wake and invaded the archipelago—this time to great success.

With the second coming of the Crown, we were banished from Eden. The Earth's sacred spell broken. Sins of nature became one with sins of the flesh.

In Cebu, the wife of Rajah Humabon fell under the spell of a statue of the Baby Jesus. She wept and begged to be baptized.

These sins were like soil that required a cleansing.

Beyond Cebu, Legazpi continued what Magellan began and cleansed kingdoms that would not bow to Christ and Crown.

I think of how babies being baptized look as if they're being drowned.

Conquistadors christened the archipelago "the Philippines" after King Philip II of Spain.

I was christened, took Holy Communion, confirmed and named Cecilia, the patron saint of music.

The Crown brought a regime of
repression that followed Spanish
feudalism and Catholic patronage.

Instead of dancing, I learned to
repress my growing body. I knew
how to follow: when to kneel and
when to stand, when to bow my
head and when to perform the
sign of the cross.

Spanish friars clamped down
on revolts for liberation and
broke up the natural development
of growing barangay society.

I even married in the church,
seeking redemption in liberation
theology. But in truth, I had a
Catholic ceremony out of habit
and to keep the family peace.

Spanish friars celebrated the
"peaceful" conversion of natives,
whom they judged to be "docile"
and "simple."

Eventually I would learn this too:
how to celebrate judgment of others,
how to be docile and suppress desire,
how to simply lock my hands and
close my legs, even as a child who
wanted to play in the pew.

They praised how Filipinos received
Christ with "eloquent enthusiasm."
By 1550, half the population of
the Philippine archipelago bowed
to Christianity.

We bow but we do not surrender.
We wield a double-edged sword,
carry the crown of thorns like an
heirloom, grow blooms among
the thorns.

The people of the islands—once
Magat, once Lakandula, once
Zula—became a Hispanicized
body, now Ayala and Bonifacio
and Buenaflor.

We flip the script with eloquence.
We have our own trinity to remember:
in the name of Maria Magdalena,
the Santo Niño, the Black Nazarene.

The Body of Christ.

Amen. This is the Body of Christ—
black in skin, devoted to the suffering of laborers and the poor. This is
his best apostle, Maria Magdalena,
the patron saint of Philippine Revolution.

The Blood of Christ.

Amen. This is the Blood of Christ—
the same blood that coursed through
the Santo Niño, the baby transub-
stantiated into Diwata—made water
god to bring rain when we dipped
him in the sea.

THROAT

Our throats hold an archipelago of languages.

The Crown considered the Philippines a way station for imperial trade. They did not waste time teaching Spanish to the Indios.

We were meant to be mute, converted into bodies of war to fight the Dutch, Portuguese, British, any who would threaten the Crown's choke hold on the Spice Islands and Silk Road.

Spanish friars found it easier to convert locals, and also to keep them divided, in their native tongues.

Yet we appropriated the king's tongue, divided it to pieces, mixed it with our lasa, seasoned and aged over three hundred years.

Savage Filipinos resisted Spanish attempts to stay silent and divided, the Christianized parts of the archipelago grew into a cohesive body that would agitate to become a nation.

By 1896, revolts grew into revolution, launched by the infamous Cry of Balintawak.

Cohesive bodies are designed to fight back.

We did not remain silent. Though the water washed over our heads, we opened our throats to cry for freedom—the freedom that Jesus would surely know we deserved.

On August 26, 1896, a group of revolutionaries from the Katipunan secret society tore their cedulas and cried, "Long live the Philippines!"

"Mabuhay ang Pilipinas! Mabuhay ang Pagsasarili!"

From the far reaches of Palawan to Bataan and Ilocos Sur, the people of the Philippines rose up in insurrection.

We are kin to those who cried out at the Grito de Dolores in Mexico, the Grito de Yara in Cuba, the Grito de Lares in Puerto Rico. We are part of a history of throats opened to revolt against King and Crown.

Meanwhile, across the Pacific Ocean brewed a new electric storm.

We are also kin to those burned by the ambitions of late imperialist dreams.

Teddy Roosevelt, assistant secretary of the US Navy, charged to control the water. The Earth's land was already carved by competing empires, but the seas—he could come for the seas.

Our bards would sing a common song of islands gilded by that line where sand meets sea.

In 1897, Roosevelt said "I should
welcome almost any war, for I think
this country needs one."

In the age of empire, our coastline
was pirates' treasure, the type of
bounty that would start wars.

America set its sights on looting
Spain's dying empire, and so the
Spanish-American War began.

And so the Spanish-American-
Cuban-Filipino War began.
If we were made kin by coastline,
by rulers and rebellions, should we
not continue to open our throats
to cry for common cause?

While Spaniards were honest about
their brutality, the Americans were
two-faced about their treachery.

We can be honest about this: we
did not live in silence under Spain,
nor did we go mutely into the
American night.

Filipino revolutionaries paid Americans for arms to fight the Spaniards,
and the money changed hands but
the arms never appeared.

They wrap their arms around us in
headlocks, throw their voices like
ventriloquists, until we say that
America was not our cheater,
but our liberator. Until we say
that truly, the location of America
is in our hearts.

The United States declared war against Spain on April 25, 1898, and on May 1, 1898, the US navy destroyed the entire Spanish fleet in Manila Bay—the only US fatality was due to a heart attack.

Our ancestors knew the truth: our gritos were fatal. It was the power of our cries that electrocuted the Crown.

The news electrified Americans. "Almost everywhere in the United States the rustling of the pages of geography books could be heard, for they were ignorant of the location of the Philippines."

We were ignorant of the Americans' end game. We were willing to believe in a gentlemen's agreement: to unite with America to win freedom from Spain. But there were no gentlemen at the heads of these empires.

The Spanish governor-general declared that he was "willing to surrender to white people but never to n———s."

May we never surrender this truth: Spain sold the Philippines to America for twenty million dollars, and America sold out Filipino soldiers for their own imperial gain.

The two empires made a deal: stage a mock-battle with America as the victors, without the Filipino troops who gave life and limb to topple the Crown.

Our ancestors' throats and limbs, their cries for independence and struggle for self-governance, all struck down through a closed-door deal.

"One minute after the Spanish flag came down over Manila, an enormous US flag climbed the flagpole in its place. The band struck up 'The Star-Spangled Banner.'"

Our fortunes were again cast with new colonizers, our gritos drowned by bombs bursting in air. But in our throats, brave cries for freedom still ring.

BACK

In US-occupied Manila on February 4, 1898, an American soldier shot three Filipino soldiers.

We shot back.

American clapback was swift: they took advantage of the skirmish. The eagle spread its wings to block the Philippine sun.

Wars are always beginning in our history. Our independence continuously eclipsed.

US troops were ordered to move through Philippine lines, and the Philippine-American War began.

In our history, there is a fine line between war and peace. We are descended from battle, from what a relative has called "war on repeat."

This war has been called "America's Forgotten War." Scholar Luzviminda Francisco calls it "America's first Vietnam."

These wars are forgotten by many, unrecorded in collective consciousness, but inked on brown bodies like invisible tattoos.

On the very day the war broke out, Rudyard Kipling published his famous poem, addressed to the United States, regarding the Philippines:

Take up the White Man's burden—
Send forth the best ye breed—
Go, bind your sons to exile
To serve your captives' need;
To wait, in heavy harness,
On fluttered folk and wild—
Your new-caught, sullen peoples,
Half-devil and half-child.

Our brown bodies are burdened by captors with delusions of white grandeur. We sense this even today. Frantz Fanon wrote, "The defenses of the colonized are tuned like anxious antennae waiting to pick up the hostile signals of a racially divided world."

American troops called the first battle of the Philippine-American War a "quail shoot." American soldiers and imperialists called Filipinos "n———s," "barbarians," and "savages."

We call on our collective history of survival, we summon our sixth sense of defense to detect imperial barbarians masquerading as friends.

In April 1899, US General Shafter declared, "It may be necessary to kill half of the Filipinos in order that the remaining half of the population may be advanced to a higher plane of life than their present semi-barbarous state affords."

We've been treated as half human, declared worthy of extermination, shot for elicitation of sexual urges, beat down on the way to church, hit on the head 125 times.

"The supremacy of the United States must and will be enforced throughout every part of the Archipelago, and those who resist it can accomplish no end other than their own ruin."

Our savage supremacy of defense is how we resist ruin. This defense, a gift from our ancestors, flows like molten gold in our nerves.

Admiral Dewey "steamed up the Pasig River and fired 500-pound shells into the Filipino trenches at close range with pulverizing effectiveness. . . . dead Filipinos were piled so high that the Americans used the bodies for breastworks."

Our bodies have been ballast on the banks. Our bodies have been silt in the riverbed. Our bodies have been trophies of American conquest overseas.

An American soldier wrote home, "On Thursday, March 29th . . . eighteen of my company killed seventy-five n———r bolomen and ten of the n———r gunners. . . . When we find one who is not dead, we have bayonets."

And still we live.

The sun beat back the eagle as fiercely as they fought when they had removed the Crown. They had the light of guerrilla warfare on their side.

I have a sun tattooed in the center of my back.

For centuries, Filipino guerrilla warfare has succeeded through the support of the people.

My friend tells me it looks like a target.

The people fed and protected the fighters, who in turn protected them. Guerrillas would immediately disband and blend into the villages as peasants, leaving American soldiers with no clear target.

It is a reminder that I do not always have to fight back, to put my aggression on the back burner, to turn warlike anger down to simmer from a boil.

Arthur MacArthur Jr., who was then military governor of the Philippines, wrote about guerrilla warfare, "The success of this unique system of war depends upon almost complete unity of action of the entire native population."

But anger is also righteous fuel that can unify and escalate a people to action.

Americans escalated their brutal campaign of pacification: they now considered every Filipino an enemy combatant; there were no more civilians.

We civilian combatants have learned to look for the whites of eyes to detect signs of deceit: a retreat of gaze to the back of the head, a break in concentration.

The Americans set up concentration camps, whole villages were burned, storehouses and crops destroyed.

We are seasoned in this sense, even brilliant; burnt, swept, disturbed, our collective nervous systems have been honed for survival.

An American congressman said, "You never hear of any disturbances in Northern Luzon because there isn't anybody there to rebel. . . . Our soldiers took no prisoners, they kept no records; they simply swept the country and wherever and whenever they could get hold of a Filipino they killed him."

Our bodies are the records. Do not disturb the memories of our fallen ancestors. Many of our fathers and grandfathers and great-grandfathers, and some of our matriarchs as well, did indeed rebel, and won the fight against extermination.

These battles of extermination extended from north to south and included the Balangiga massacre of 1901 and the Bud Dajo massacre of nearly one thousand Moros in 1906.

We retain their fierce dedication to survival, and extend the fight to the battleground of love.

Immerwahr calls these massacres extensions of massacres at Wounded Knee, Sand Creek, and Bloody Island, yet the massacre at Bud Dajo "dwarfed them all."

How to move beyond the massacre of bodies, and the massacre of trust? Better to prolong the war, especially among ourselves.

The Philippine-American War went on until 1913, spanning the course of fourteen years. Immerwahr notes that after Afghanistan, the Philippine-American War is the United States's longest war.

We don't have to have wars that span generations. We can have one another's backs.

General Bell estimated that more than six hundred thousand Filipinos in Luzon alone were killed or died of disease as a result of the war.

We don't have to be collateral damage. The divide and conquer of cancer, the ambush of heart attacks, the attrition of diabetes. Our bodies continue to bear the long burden of disease, and still we endure.

But Luzviminda Francisco enumerates that more than one million Filipinos died during this forgotten war.

When we are met with erasure, we story back with permanence. On my back is the word "pagmamahal." "Big love" in Tagalog, needled in permanent ink.

American blockades meant Filipinos had little access to firearms. These casualties were due to superior firepower, not due to superior humanity.

We can step up to the superior path of love. We have its marks on our body.

"[Filipinos] stood up to the heavily armed Americans with spears, darts, the ubiquitous bolo, and even stones, prompting General Lawton to remark, ' . . . they are the bravest men I have ever seen.'"

We take back and flip the meaning of battle, armed with spears, darts, and stones, ubiquitous bolos of transgenerational bravery and love.

BELLY

Filipinos have a transgenerational habit of belly laughing through the worst of times.

American imperialism was an expansion from the belly of the beast—an outgrowth from Native conquest, manifest destiny, and slavery.

Like all of us descended from state-sponsored violence, we grow humor in order to survive.

Here's a fact that rarely survives the spin cycle of US history: the Philippines was brought into the American fold through age-old traditions.

It is an age-old tradition among the oppressed to spin kwentos and tell jokes, to wield laughter like a weapon to survive.

United States imperial expansion required the soft weapon of appearing magnanimous to the outside world as a matter of foreign policy.

Laughter is a magnanimous gesture to ourselves, an offensive maneuver to show that they cannot kill our joy.

Americans sought to demonstrate their benevolence by not killing too many Filipinos in public and by building health-and-welfare programs, investing in public infrastructure, and redistributing land held by Catholic friars.

The kind of joy that emanates from our guts is a redistribution of resilience. But it should not be mistaken for benevolence. Filipino revolutionaries probably laughed at the American narrative of "little brown brothers." Maybe they told jokes while sharpening their swords.

The brutalities of fourteen years of the Philippine-American War had no place in this narrative of benevolence, and so they were forgotten (translation: erased).

We have not forgotten that, as comedian Rex Navarrete says, Filipinos are like chocolate-covered pretzels (translation: brown on the outside and twisted on the inside).

This narrative of "American benevolence" would be twisted into one of "American savior" as a new kid showed up on the imperial block.

It's funny how many have come for us. New imperialists to replace the old. We are so wanted. We have so much to desire.

"Asia for Asians" became the new imperialists' slogan, when the truth was, they wanted Asia for Japan.

"The Philippines for Filipinos" could have been a slogan for the resistance, but guerrilla fighters are more poetic than that. The truth is, we like to laugh, but Filipinos are nobody's fool.

As early as 1906, the American government considered giving the Philippines its independence—not out of benevolence but to avoid fighting Japan.

We've been called many names, but "coward" was never one of them.

Shortly after the Japanese attacked the Philippines, American officers retreated to Australia. Most American soldiers were captured or ordered to surrender.

We were conquered and abandoned, but we never surrendered. And now we cannot be avoided. The joke's on them—the gallivanting Spanish, the retreating Americans, the overzealous Japanese—because look at us: we are not only independent, we are everywhere. We are our own diasporic empire.

During the Bataan Death March, soldiers sang:

We're the battling bastards of Bataan;
No mama, no papa, no Uncle Sam.
No aunts, no uncles, no cousins, no
nieces,
No pills, no planes, no artillery pieces
And nobody gives a damn.

We are fighting Filipinos who will always fight for freedom; we are karaoke-singing bastards who will always give a damn.

Within two months of the Japanese occupation, one Filipino writer wrote:

"It was as if the Philippines had become one vast military prison. . . . Every day on my way to the office, I run across dozens of Filipinos who have been tied to posts as punishment for some trivial offense which they have committed. Usually the victims are black and blue or bleeding from the terrific lashings they have received."

And when times get terrible, we got jokes. Dark ones. Terrific ones. Black and blue and bleeding ones. We always. Got. Jokes.

"No one, during the darkest days of the occupation, could sleep soundly. Everyone was waiting for the hour he would be arrested and tortured."

An internet meme on a brief history of the Philippines shows the country as a ball, arrested and chained to its various invaders.

Guerrilla units of resistance once again formed all over the island to break the chains of Japanese rule.

With every new colonizer, the Filipino ball gets a new makeover, a new Band-Aid, or a valiant new scar.

The guerillas fought valiantly, perhaps seasoned by the scars of experience passed on from previous generations.

In the end, the Filipino ball apparently organizes and becomes free to reexperience what it once was and will always be, a gut-deep identity inherited from previous generations.

US war planners wanted General Douglas MacArthur (son of Arthur) and his troops to leave the Philippines to its own devices and to instead take back Taiwan as a way to conquer Japan.

Left to our own devices, free of colonial conquerors, the Filipino ball becomes balut. The essence of Filipino identity.
Balut, balut, balut.

But MacArthur was determined to keep his promise of "rescuing" the Philippines.

"Laughing saves the family," my aunt used to say. "We have a lot of problems, and for every problem we have a joke."

In the end, this "rescue" caused more suffering for Filipinos than MacArthur had intended.

My aunt knew in her gut what Freud had to write a book about to understand—that jokes take back pleasure otherwise surrounded by suffering.

When MacArthur returned to the Philippines in October 1944, his troops surrounded Manila to take back the capital.

My uncle also understood that humor was a way to take back power and triumph over indignity.

Though history celebrates this moment as the triumphant bookend to MacArthur's declaration, "I shall return!" it was more a moment of "I shall blunder!" MacArthur had cut off the Japanese Army's escape route, prompting what would become one of the bloodiest battles of the Second World War.

My uncle once waited all day in a line that surrounded the US embassy in Manila, but they rejected his visa a third time. With his route to America cut off, he picked a fight with the embassy officials.

Japanese soldiers behaved like cornered animals. They blew up factories and warehouses and took down power and water systems. Large hotels, including the Manila Hotel, became sites of organized mass rapes. Whole families were slaughtered, babies were bayoneted, pregnant women disemboweled.

An embassy officer called him a monkey, and so my uncle began to behave like one. He yelled, "Oooh oooh aaah aaah!" and scratched his armpits. He waved his hands in the air, but the best was yet to come.

This does not even count the American shellings and bombings that were to come.

While the embassy officer watched my uncle's hands wave, my uncle took advantage of the confusion to pop a heavy punch to the officer's jaw, then ran away.

The 37th Infantry Division led most of the American combat in Manila, and it was known for being heavy-handed with its artillery. This was part of their strategy of saving American lives.

Our strategy: everyday acts of resistance. Turn collateral damage into offal stew. Transubstantiate the belly of the beast. Reconstitute its tripa as the bomb dinuguan and power pinapaitan of rising again.

Americans bombed the Philippine General Hospital, which housed more than seven thousand Filipino civilians. This was just one of many refuge centers that became collateral damage during the rescue effort.

This is how we gut the monster. Clean its entrails. Eat it from the inside out until there is nothing left.

The American high commissioner later admitted, "We levelled entire cities with our bombs and shell fire. . . . We destroyed roads, public buildings, and bridges. We razed sugar mills and factories . . . [in the end] there was nothing left."

The humor of the oppressed is raising something out of nothing, a peso out of fifteen centavos, multiplying survival by the factor of a Benjamin.

After the Battle of Manila, about one thousand American soldiers died and at least one hundred thousand Manila residents were killed, many buried in unmarked graves.

In the end, we can only touch into levity once we exhume our ghosts from their million graves.

Filipino deaths across the archipelago over the course of the war totaled more than 1.6 million.

For this they also try to destroy: not just our joy, but the glory of our grief.

Immerwahr calls it "by far the most destructive event ever to take place on US soil."

So from the soil of our bellies let laughter flow like groundwater trickled with sobs. Our smiles will not be minstrel shows for the master.

A photo of Manila at the end of the war shows a city razed to the ground.

My grandfather was a prisoner of war. The Japanese threw him to the ground, filled his belly with water, stomped on his stomach to let the water go. At the end of the war, his body was never found.

The remains of buildings jutted from the earth like jagged teeth of exhumed jawbones.

Did my grandfather's body lie in that wreckage? How many of our kababayan were similarly reduced to remains?

After World War II, Manila was one of the most devastated cities of the Allied front, second only to Warsaw.

In the face of devastation, we laugh to not cry.

Warsaw was painstakingly rebuilt, in part by the Soviet Union. Meanwhile, the US focused on rebuilding Western Europe, but there was no Marshall Plan for the Pacific.

Perhaps laughter is also a form of mourning.

The US did grant some postwar aid to the Philippines—with purse strings attached.

To laugh from our bellies is to pull on our heartstrings. To pull on our heartstrings is self-repair and self-aid.

In exchange for postwar aid to repair damage, the United States retained the right to extract natural resources—including human labor—from the islands, and the neocolonial period of Philippine-US relations began.

We are never quite posttraumatic, never quite postcolonial. Like my uncle, we continue to punch up and persist until we have the last laugh.

FEET

As a child, I used to walk on tip toes to avoid contact with the earth. Now, I let my flat feet sink ungracefully to the ground.

I had been taught the ground was soiled, and, therefore, a threat akin to sin. Now I know that soil is sacred territory.

And so I plant my soles, and the rest of my body contorts to adapt. Along with the pain of walking differently comes great relief: I no longer have to resist gravity. Though we may fight our fellow humans, the earth is designed to support us.

"We're tired," says rapper Rocky Rivera. "We've been resisting for so long."

And so some say: "Walk away from what's come before. History is heavy. History should be left in the past."

But we carry history with us whether we face it or not. James Baldwin writes that we must wrestle with history, contest with it, in order to bring ourselves out of it.

My body concurs. Its lightning bolts and embers remind me that history works its forces on us all. Many of our bodies long to read the silent scripts buried in our nerves.

And so we dig up history, and we work it. We flip the conquered word and reverse it.

And when we do, we find pain, but we also find transcendence. Histories reveal that we are deities of survival. We are the "goddesses of guerrilla warfare in every lifetime," the great-grandchildren of babaylan who swam with crocodiles, the granddaughters of revolutionaries uncaptured and witches who would not be burned. The more we root soles in this history, the farther we can stretch crowns to the sky.

I have a long way to go before I can move with ease in this marked body, this enduring shell of ancestral pasts. But I know this is a first step.

Knowing history is not just knowing self, it is knowing us. It is knowing what we have collectively survived and what we, together, might become.

III
Neuroregulation

- In therapeutic terms, the nervous system's ability to move fluidly between parasympathetic and sympathetic states of rest and activation. Also, the collective work of maintaining bodily functions like metabolism, heart rate, and breathing.
- Where my healing journey begins, from ritual and collective stewardship to harmonizing with the nervous systems of others.

Not even the pain of colonization can weigh us down so that we cannot sing and dance.

—LENY MENDOZA STROBEL

Bayanihan

I approach the cliff's edge ready to give in to gravity. I believe my body will fall well, unsupported as it has been, and burdened as it is by growing pain. To finally free my bones from the fight to stay undrowned—this seems like a logical choice, and a merciful one.

It is October 2001, three years since I relocated to San Francisco and six weeks after 9/11. George W. Bush has just launched Operation Enduring Freedom, invading Afghanistan and eventually sending troops to the Southern Philippines. The aura of tragedy hangs thick in the air.

I am twenty-five and living with pain so debilitating that I've quit a nonprofit job working with the Bay Area Filipino community. I don't yet know to view my pain as an illness. I simply think of it as a personal failure. And because I have seen multiple doctors who are unwilling or unable to help, my only treatment is a handful of Aleve tablets every day.

Even more debilitating than the physical pain is emotional pain that has been rising slowly like the tide. In 1999, I came to San Francisco to find a place I could call home. After a lonely upbringing in the Midwest and a difficult college experience in the Northeast, I am finally building community closer to ancestral shores. Prior to my move, I tried and failed to bond with other Filipino American college students in the Boston area, who organized thong contests and mocked student protests for ethnic studies.

By contrast, in San Francisco, after about a year of trial and error, I found a community aligned with my values. I'm now a grassroots Filipino activist, which feels like a homecoming. I joined Diskarte Namin, a band that has been the heart of my new life. I work two jobs: one at the Westfield Mall selling pens by day, and the other at a Top of the Hill karaoke joint serving beer and pulutan by night. And yet my body is alone in dark waters. I feel increasingly alienated from others.

For months I have been relying on a Post-it note to get me up in the morning. The note is dull yellow with an outline of an eight-rayed sun. Inside the sun, I wrote: "Good morning! Live in the world with joy." Every day I look up at the note and heave out of bed. I brush my teeth and examine it again. This morning I read the Post-it and searched my body for joy. I found none. There was only a dull sense of dread, a vise grip in my sacrum, sea glass in my wrist, and an overwhelming feeling of exhaustion—the kind of fatigue that hurts. I made the decision there and then to act on a plan that had long been materializing in my mind.

The morning was glorious. I got in the car and drove from my Mission apartment toward Fort Funston, an attractive park of sandy bluffs that fall into the Pacific Ocean. A rare glare of sun burned the fog that normally settles thick in this region. The only trace of fog today is inside my brain, and it is unsettled like a hurricane's winds.

I thought the bluff would be crumbly like a shortbread cookie and that I would slip and fall on a slide of bone-colored California clay. But its edge is both soft and strong. Soft because low-growing succulents enrobe the ground and pad my footsteps. Strong because, as I bend my knees to test a small hop, the cliff seems to push back at me—a solid accomplice. I am not so unwavering. Can I really take this leap?

I had considered jumping from the Golden Gate Bridge, but I didn't want to end up submerged in water and carried away to sea. Fort Funston seemed the perfect alternative. The beach would be wide enough to keep my body away from the tides. I want my body to be easily found.

Earlier that year, I went to a coworker's wedding and was disturbed by their vows. The groom said to the bride, "I love how you wake up joyful and ready to tackle each day." His declaration was earnest and sweet, yet it stuck with me like a nettle in my heel. I wanted to shake it off, but instead I looked closely at it barbed in my flesh. *Do people really wake up that way?* I asked myself. What was wrong with me, that I had no experience of facing each day with joy?

I thought everyone woke up with dread and fear and a fifty-pound dumbbell dropped on their chests. It took inordinate energy to cast that weight off and swing my lead legs out of bed. Yet lead and dumbbells are not exactly the right metaphors. I didn't feel numb and unmotivated like deadweight; I felt agitated and excited and terribly afraid. If I didn't muscle my way out of bed every day, I might spontaneously combust and turn to ash between the sheets. I was driven by an unwillingness to surrender to annihilation. I was also driven by the need to pretend: pretend that I was normal, pretend that I actually belonged in the realm of people who belly laughed and threw their heads back with ease, pretend that I could be fueled by joy instead of by the iron prod of shame.

Today I am tired of pretending. My skin hurts from containing the jagged, broken pieces of me. My nerves are too tender, singed, and burnt out. I want to be carried away by a higher power. I am afraid of heights, and it's because of this that I chose gravity as a method for ending my shame and my pain. Maybe in the free fall I might even glimpse a moment of joy.

From the edge of the cliff, I look to the horizon. Just then, a hang glider whispers by, no more than twenty feet away from where I stand. I stumble backward, startled by the person in the glider who seems to look me right in the eye. That brief moment of connection is a jolt to my nervous system. It is only then that I realize the practicalities of my situation.

I tiptoe back to the edge, steadying myself against vertigo as a wave of dizziness and nausea washes through me. I look down and see that the bluff is steep, but not steep enough. The sand below is soft. In my mind, the cliff was much taller and the sand more punctuated by hard, protruding rock. If I jumped, I might hit a couple edges on the way down but end up alive, with a broken ankle or limb. *I could go headfirst*, I think. I could aim to break my neck on impact, but what if I failed? It would only prove to others the embarrassment I know myself to be and land me in even more excruciating pain.

I taste salt on my lips as tears wet my face. All I can think now is that I have to get back to the car. I return to pretending. I don't want anyone to know what I was attempting. They would see I was stupid. They would laugh at my weak idea of suicide and rightly conclude that I couldn't even do this one last thing right.

I manage to drive myself home under a crushing cloud of defeat. Back in my apartment, I climb into bed and fall asleep at the height of the sun-kissed day.

Alone in the Flood Zone

I have long wondered what brought me to this moment of despair. I had weathered twenty-five years of sadness and not once in all that time had I tried to act on suicidal ideation. Why had I almost taken my life when I was finally building the home I wanted?

People who knew me at the time have asked what I was feeling, since they had no idea that I was suicidal. If I had to describe it in one sentence, it would be this: I felt that every day I stood alone in a flood zone, facing roiling waves that threatened to wash me off the sand. The community around me was enough to help me face these emotions. But it was not yet enough to keep me from the cliff's edge.

There is a special agony to feeling alone when surrounded by people. In college I was among an expected community of peers. And yet most of my days, while other students milled around bantering, I wore a solitary path between my dorm and the science center, head down, saying no more than "hi" to one or two people for days. From cold November evenings when snow dusted the ground to blistering June days when sun wilted the ivy, this was more often than not my singular routine. I was depressed and my chronic pain was intensifying. And yet not once during my five years in college did I attempt to take my life.

Meanwhile, in San Francisco, surrounded by a web of people who were growing to be my community of peers, I began to understand that I truly wasn't alone. This feeling, at first, was painful. It was like the inevitable rise of the tide: an organ-deep ache, slow and creeping, unconfined. When I was more fully cut off from others, I was also better able to compartmentalize the pain, to titrate between tingling, shooting, aching, and an analgesic numbness that was a necessary relief. Now that I was more open, the pain threatened to dissolve my coping mechanisms, to breach all barriers of defense.

Becoming part of community I cared about also brought new risks and anxieties, as well as a somatic understanding of what I had been missing all along. This hit me like a slow flood of grief. The waters rose to engulf my body until I felt I could float no more.

In public, I dug myself into the ground to stand tall and get things done. In private, I allowed myself to be swept away. Any experiences of growth and enjoyment in the day would fall away, subsumed in a tsunami of anxiety-driven thoughts: *If I disappear, no one will notice; I am useless; I am exhausted; I am a mistake.*

I have told people who ask that they couldn't have known I was suicidal because I worked so hard to hide it. The day before I found myself at the edge of the cliff, I had been at a birthday party for someone who remains a dear friend. I walked around her apartment giving people hugs and daps. Thick weed smoke filled the rooms. Though I made small talk with acquaintances and laughed at the jokes people cracked as they passed around joints and blunts, I was also watching everything as if floating from afar. I saw the birthday girl double over with laughter while another friend slapped her chummily on the back. I saw a group huddle together to shield a flame to light the bowl cupped in someone's hand. And while I positioned my body in poses of relaxation, I suppressed a voice inside that said, "Get the fuck out before you do something to make them hate you."

I had spent my whole life living with pain and fear while pretending I was okay. High functionality was my best coping mechanism. It was also my best performance. I was a body getting things done while my soul was torn open at sea.

The Commons of Being Alone

People struggling with suicidality are often told that we should remember we are not alone. My colleague Marina has shared that she bristles against this notion. She lives with serious mental illness and has attempted suicide eight times over the course of decades. Each time, she describes being in a place of despair that is actually a state of being alone, because she knows that others around her are not experiencing what she is feeling. Our interior world of mental health is distinct for each of us—no one can fully understand what the other is going through, and yet there are commonalities to the origins and depths of our wounds. There are also commonalities to how we cope.

Marina was in her midtwenties the last time she attempted suicide. She overdosed on medication, tried to slit her wrists, and got hospitalized against her will. She tells me this matter-of-factly because she believes that we should all talk about suicidality more, and perhaps because as a scientist, she is used to sharing facts in an evenhanded way. When she was hospitalized, Marina called one friend but made sure that few other people knew about her mental health struggles, much less her suicide attempts or hospitalizations. At that moment, after being admitted, all she could think about was getting out of the hospital and home to Thanksgiving dinner with her family, so they would never have to find out.

Hiding our struggles is a common coping mechanism, especially in communities where mental illness is stigmatized and poorly understood. My friend Alexis testifies to this as well. Alexis was only twelve when she first thought about ending it all. She had been carrying a burden of loss since infancy. Her mother died from cancer when Alexis was only seventeen months old. She lived with this grief along with her father's gendered expectations that she carry the load of domestic duties for the family.

One day, after Alexis's father gave her a long list of chores, she felt so overwhelmed that she grabbed a bottle of bleach, unscrewed the cap, and brought it to her lips. Her brother saw her and stopped her before she drank a drop. That evening, Alexis had her first panic attack. Overcome by shame at having been caught, she wouldn't speak of this suicide attempt, or her other struggles with mental health, for years to come.

Imee, with whom I've studied the Filipino martial art of Kali, experienced suicidal ideation through her teenage years and into young adulthood. Her family and friends knew that she thought about death because she spoke about it often but only in jest. She would crack jokes about dying, then laugh as a way to release steam. In this way, she may have been an example of what Kevin Nadal calls "smiling depression." Beyond the jokes, she became increasingly quiet and never acted out, which earned her praise for being "good" and "mature." All the while, on the inside, she felt hurt, increasingly isolated, and misunderstood.

Marina, Alexis, Imee, and I are just four examples of the commons of feeling alone. We are also examples of a disturbing trend among Filipina Americans. A Centers for Disease Control study has shown that young Filipina Americans experience the highest rates of suicidal ideation compared to others in their age group, almost double the rate of young white women and young African American women, and more than 10 percent higher than young Hispanic women. Young Filipina Americans also exhibited the highest rate of suicide attempts of both males and females in the study. Ironically, the feeling of aloneness that can contribute to suicidality may be exacerbated by the Filipino value of pakikisama, or social acceptance, in which pressures to keep group harmony can trump the urge to seek help.

The hope for me lies here: the commons of being alone is matched by the common ground through which we heal. The four of us share the belief that culture, community, and ritual are necessary to keeping us grounded in this world. The ways we find these healing paths are as varied as the currents in the sea, but they all start with finding a sense of home.

Homing Instinct

In precolonial times, the people who inhabited the Philippine archipelago lived in homes made of the plants that constituted their environments. These homes were called bahay kubos, which literally means "cubed homes." They were also called "nipa huts" since their thatched roofs were made from the broad leaves of the nipa palm.

In the Western fable of "The Three Little Pigs," a thatch-roofed hut is a symbol of insecurity and naiveté. The Big Bad Wolf easily huffed, puffed, and blew over the first pig's frail home. In contrast, the red brick house of the third pig is a symbol of strength and impermeability. But in places like the rural Philippines of olden times, integration with the environment was a foundation of strength. Nipa huts were embodiments of this integration and of the resilience of permeable and portable housing.

As part of our healing journeys, Marina, Alexis, Imee, and I have all been looking for our nipa huts in some way, our own little corner of the world in which to shelter in community.

Marina lives in the United States, but her father repatriated to the Philippines. He repeatedly asks her when she is coming home. When he asks her this, Marina says she feels a deep longing in her bones. She has come to realize that part of her depression is due to being separated from her homeland.

Imee grew up in Ohio, but she left home at eighteen due to tension with her parents. She learned to not only survive but to thrive on her own, presenting herself as capable, reliable, and organized—"Like I had all my shit together," she says. But she remembers screaming to herself on the inside, constantly saying in silence, "I want to go home, I want go home," only she didn't know where home was.

I grew up in the isolated confines of a red brick house on a hill in southwest Chicagoland. In that impermeable environment, I was exposed only to small clues toward a community to which I might someday belong. My parents spoke Tagalog in whispers, which grew to more animated levels the few times my grandmother Lola Simeona came to visit. When Lola opened her suitcase, I would rush over to inhale the smells of tropical humidity and sweet carabao milk candy that emanated from within, scents that imprinted me with longing and hope.

In 1989, I visited the Philippines for the first time as a teenager. This experience showed me that my own individual body was connected to a greater body of ancestry. I didn't yet think of the islands as home, but I did think of the Philippines as a part of me that I wanted to get to know—the place from which the carabao milk candy came and where I supposedly had a village of cousins all over Luzon.

So when I was old enough to choose for myself, I moved closer to the Philippines, to California. Like salmon on the verge of spawning and death, I followed the magnetic pull of a natal source and migrated upstream.

The Necessity of Ritual

In naturopathic work, there is something called a "healing crisis," in which you start to get much worse before you get better. Imee calls it a "healing spiral," Alexis calls it a "nonlinear process," and Marina calls it a "tidal wave." My suicide attempt was part of this nonlinear healing journey, this spiral, this wave, and as a crisis, it presented new healing opportunities. I'm just thankful I had community to catch me before I drowned.

When I returned home from Fort Funston the day that I almost jumped, I slept for hours. At one point, I heard my roommate come and go, but I stayed silent and pretended I wasn't home. In the evening,

I awoke to the doorbell clanging. I peeked out the window and saw BJ, one of the first real friends I had made in the Filipino community in the Bay. He caught me peering through the curtains and dug his hands into his messenger bag, surfacing with a burrito and a joint. A large part of me was relieved to see him, but another part wanted to keep hiding, to avoid contact at all costs. I didn't want to talk about what had happened; I was too ashamed to admit that I had tried—and failed—to take my own life. But I was hungry, and at heart, I was longing for connection, and so I opened the door.

Looking back, I believe the simple interaction that followed was a beginning toward feeling truly at home. BJ proceeded straight to the living room, where he tossed the joint onto an end table and sat cross-legged on the floor. While he unwrapped one half of the burrito, I sat perched on the edge of the couch, watching him move through a slow-motion filter in my mind. He prattled on about a Filipino culture show he had just done at Sacramento State, and as he talked, I felt my body begin to relax into the sofa. I hadn't even realized I was braced at the edge of the cushion just as I had braced my body at the edge of the cliff.

BJ was almost done eating by the time he realized I hadn't even made a move toward my food, so he put his burrito down and threw me the other half as if throwing a dog a bone. I caught it, unwrapped it, and devoured it in four big bites. I hadn't said a word about what had happened, but I imagine that my puffy red eyes and defeated demeanor showed him that I needed help. "You okay?" he asked. "Yeah," I responded. He accepted that and stayed, and we smoked the J and watched infomercials all night long.

Hanging out with BJ was like constructing my first nipa hut. If a healing crisis had led me to the edge of the cliff that day, a new healing shelter kept me from going back. From that evening forward, I began dealing with my pain not in isolation but in relationship with others like BJ, with whom I was beginning to feel safe.

I also see my suicide attempt as a grasping at ritual. Rituals are often designed to reduce uncertainty, invoke control, and alleviate pain and grief. In this way, my suicide attempt was meant to bring a certain and controlled end to my pain, when I felt I had no other choice. But it was a ritual of solitude which, if successful, presented no option of repetition toward healing or spiritual elevation. In this way, it was a poor substitute for what I really lacked: a communal practice for processing pain.

While writing this book, I learned from scholar Karina Walters about the Choctaw re-walking of the Trail of Tears. This is an annual remembrance of the first US Indian removal program, one of the most egregious land grabs and displacements of a people in American history. Over the course of three brutal winters from 1831–1834, more than eleven thousand Choctaw people were forced to walk 550 miles at gunpoint from their homeland in what is now the Southeastern United States to reservation land in Oklahoma. It was the beginning of decades of disruption of Choctaw lifeways as the US government severed people from their homes, their culture, their language, and their connection to original instructions. The impacts would last for generations to come.

During the re-walking of the Choctaw Trail of Tears, participants tell stories about ancestors and their forced migration in relationship to landmarks along the way. Around the same time that I learned this, I found out more about the landscape of colonization imprinted in the San Francisco Bay Area. It turns out that the cliff I wanted to jump from—indeed, the whole span of Fort Funston parkland and shore—was named after an American general credited with maintaining American rule in the Philippines.

Frederick N. Funston, also known as Fighting Fred, earned his name by leading the 1901 mission to capture Filipino rebel leader Emilio Aguinaldo. In 1906, Funston became commander of the Presidio, the base on the northwestern edge of San Francisco from which the majority of American soldiers was deployed to fight in the Philippine-American War. I had come to the Bay to be closer to ancestral shores, yet I didn't know till years later that these shores were so deeply implicated in the conquest of my people. This led me to wonder, *How do we as a community process the generations of trauma embedded in our bodies and also in the land around us?*

The remembrance of the Trail of Tears provides a model. Despite the destructive impacts of the original tragedy, Walters describes the annual walk as a ritual of creation. Thousands of Choctaw people re-walk a portion of this terrible journey so that younger generations will not forget their history, but also so the community can generate new practices through discussions about the future. During the walk, people point out landmarks of survival and also ask each other: *What is our medicine? What kind of ancestor do I want to be? How do we maintain balance, and when we get out of balance, how do we heal ourselves?* Walters has found that modern-day rituals like these can ease the symptoms of what she calls the "colonial trauma response," whose symptoms include depression, substance use, and suicidality.

Filipino scholars like Leny Mendoza Strobel, Kevin Nadal, and E. J. R. David taught me that Filipinos experience a similar colonial trauma response from our collective history of colonization and migration. And yet, back in the day, we had no regular rituals at the Presidio or at Fort Funston through which to re-member the imperial aggression launched against our people from these cliffs. I had heard of City College students going to visit Manzanar, a Japanese American internment site during World War II. I knew that activists took trips to Angel Island to process the traumatic confinement inflicted on Asian immigrants detained there. Opportunities to re-member and heal historical trauma were all around us. But perhaps as a community, we are re-membering and healing in other ways.

Homing Ritual

Walters's story about the Trail of Tears made me think of a traditional Filipino ritual called a "bayanihan." In the old days, just before monsoon season, villagers in rural areas would gather to move nipa huts to higher ground. Nipa huts were large enough to house entire families, and so they were quite a burden to transport despite their flexible design.

To lighten the load, Filipinos fabricated a tool to distribute the hut's weight among a group of fifteen to twenty people. The tool was a sturdy frame of bamboo poles, crisscrossed and latched together with ribbons of palm leaves. Some villagers would lift the hut off the ground, just enough so that others could slide the bamboo frame underneath. Then, together, the group would heave the hut upward until the protruding ends of the frame found support in the crooks between necks and shoulder bones. Sharing the burden of the hut among them, these villagers would walk together away from rising tides, away from low pooling places and toward more stable ground. Once the nipa hut was safely moved, the family would cook a feast and host a fiesta for the village around their newly relocated home.

This ritual is called a "bayanihan" because "bayan" means "people, town, or nation." The moniker implies that the ritual of moving homes is an enactment of how to be a nation. To my knowledge, bayanihans no longer occur except when recreated for show at town festivals, but bayanihan remains a cultural value throughout the Philippine diaspora. It is a value based in the belief in kapwa, or the self in others, which guides rituals of mutual aid and collective action today.

This led me to wonder: *What are our modern-day bayanihans?* What rituals have we created, if any, that can help alleviate depression, suicidality, and other effects of accumulated trauma in our communities?

Activism as Bayanihan

When I moved to San Francisco, I was instinctively looking for people with whom I could enact rituals of healing. I found one of them in Rachel. Rachel was a true organizer who worked for a labor union and organized young Fil-Ams in San Francisco the rest of her time. I met her in Seattle, at the 1999 protest against the World Trade Organization, which I was covering as a freelance reporter for *Mother Jones* magazine. When I introduced myself, she said, "You live in San Francisco?" The surprise was evident in her face. "What's your number? Come march with us tomorrow!" And I did.

Back in the Bay, Rachel didn't let me disappear. She called a week later and invited me to a meeting of the Committee for Human Rights in the Philippines. This was December of 1999. I was thrilled to be invited but nervous about meeting others in the group. Would I fit in? Would they accept me? Would I be too awkward to belong? When I entered the room, I saw about a dozen other young Fil-Ams sitting cross-legged in a circle on the floor. I remember taking a big swallow and pausing at the door. That circle felt incredibly intimidating. But then folks scooted back and made room for me.

And so Rachel introduced me to Filipino American activism, which, for a time at least, became a practice of great healing and growth. Over the next few years, I developed relationships both deep and wide with Fil-Am activists from LA to Seattle. Together we organized marches to support Filipino World War II veterans and airport workers, protested human rights violations in the Philippines, rallied against Muslim profiling after 9/11, and took direct action against profiteering companies when the war on terror began in Afghanistan, the Southern Philippines, and Iraq.

During these mobilizations, I felt a surge of inspiration and belonging. My body buzzed with being alive. And then I would go home and feel down and in pain again. For a while, activism was a boom-and-bust cycle, an addiction almost, which at its lowest point would lead me to that cliff at Fort Funston. But steadily afterward, the times between street actions became filled with companionship from BJ and from others in the community, and with sharing meals and making protest music. Activist events became an activist life, and each day I felt more confident, more supported, more integrated into the world. My chronic pain was not technically getting any better, but the emotional pain in my heart was starting to heal.

I've since learned that a strong sense of ethnic identity can improve symptoms of psychological distress in some Asian American communities. Studies have also shown that voting, volunteering, and activism can help ease symptoms of mood and anxiety disorders. Over the years, I have had the privilege to pursue a panoply of individual therapies, all of which have helped calm my nervous system and provided some relief from chronic pain. But it was the relational therapy of activism that buffered my pain for the long term; it became a collective ritual, a form of bayanihan that helped me move to safer ground.

Ethnic Studies as Bayanihan

For much of her adolescence and young adulthood, Alexis continued to keep her struggles mostly to herself. But then in her twenties, she began to work with the Pin@y Educational Partnerships, or PEP, a groundbreaking initiative led by Allyson Tintiangco-Cubales. Allyson and a group of her students founded PEP as a mentoring program to address the high rates of school dropouts, teen pregnancy, substance use, gang violence, and mental health challenges among Filipino American high schoolers in San Francisco. PEP educators began to use cultural tools to tell stories, employing theater and hip-hop as vehicles to teach about Filipino American history. Students were hungry for this type of identity and community building. The mentorship program grew into a yearlong ethnic studies class in several high schools throughout the city.

"This is really what has led me to this wellness journey," Alexis told me. When I interviewed her, she and her longtime partner had just relocated to a new home in Oakland, a place where they could raise their children and provide another anchor for the Filipino community throughout the Bay. While we talked on Zoom, Alexis was unpacking and settling into their new home. "This return to our story, to the inquiry to understand ourselves by better understanding the legacy of all that happened before me, to reclaim parts of ourselves that have been denied to us, this is what has healed me and led me to do the work I do today."

Alexis is now a mental health professional who supports young Filipinx Americans on their wellness journeys. When Alexis joined PEP, she was on track to become a lawyer. But then she became involved in PEP's MALONG program, which focused on the mental health and wellness of students, and she started to see her own struggles in theirs.

Alexis says that ethnic studies integrated with mental health support saved her life. Through PEP's programs she felt seen, cared for, and important. She saw her family and friends lifted up as a community of value, a community worthy of saving from rising tides.

Networked Bayanihan

Imee is a practitioner and teacher of Kali, an ancient Filipino martial art that uses bamboo sticks for hand-to-hand combat. She is also an international transracial adoptee, which she calls an experience of "being colonized twice." After a childhood of adoption, homeschooling, and the Evangelical Church, Imee longed for places where she could belong.

She found belonging in the sticks of Kali. Imee began going to the woods to practice Kali with a community of nonhuman beings—the trees and the forest spirits—touching into an animist past. She found the ritual so powerful that she decided to offer an introduction to others regardless of where they lived in the world.

Imee created what she calls "Kali Containers," online spiritual-based Kali classes to "heal the warrior within." These sessions have brought together networks of people from all over the globe, from Australia to Japan to Columbus, Ohio. In addition to those who gather in body, Imee says that participants "roll in with the spirits of their ancestors and descendants," making the online sessions a gathering of generations.

In creating this online community for others, Imee says she has found healing for herself. She says it has "spoken truth to the lie that I am alone."

Therapy as Bayanihan

Back in 2001, I had never heard anyone speak of individual trauma, much less collective trauma, historical trauma, or colonial trauma—not in my family, not in my schools, not even within the Filipino activist community.

Back then, I would not have been able to tell you what I know now: that what brought me to the edge of that cliff was pain from unbridled anxiety and depression, but also from accumulated psychological trauma. It took me forty years to understand that the traumas I had experienced were multilayered, stemming from birth through the present, extending back even further beyond. I had suspected but not accepted that I also carried historical trauma from generations past.

Back then, I had also never heard anyone speak about therapy. I had tried to see a counselor in college, but she left me crying on her couch when time was up. It took another transformational experience with activism in the Bay for me to be exposed to therapy as a ritual practice.

In 2002, Malkia Devich-Cyril and Amy Sonnie hired me as a communications trainer to work with youth organizing groups across the Bay. This brought me into what Amy calls "a twenty-first-century, multiracial, beloved community" that has shaped every bit of who I am today. Working with a phenomenal network of Black, Indigenous, Asian, and Latine and queer and trans activists, I learned principles of accountable leadership, direct and nonviolent communication, and queer and disability justice worldviews.

I also learned that it was normal to go to psychotherapy. Both Malkia and Amy scheduled work around weekly therapy. Inside our cramped office, conversations would flow between work meetings and discussing their latest therapy sessions. With therapy as part of daily talk, it became an option for me to explore. After a few false starts, I found a therapist whom I would continue to see for seven years.

For Marina, therapy has been a must. She entered the behavioral health system at a young age because of serious mental health issues, and since then, therapy has been just one part of a regimen that also includes medication, occasional hospital treatment, and interventions like electroconvulsive therapy. Together these practices constitute a sort of ritual of professional support for her overall well-being.

Marina says that despite her extensive experience with both Filipino culture and social welfare, she often felt her mental health challenges were un-Filipino, because she never encountered other Filipinos in the behavioral health system. This is a problem that Alexis has dedicated her life to addressing. Alexis started seeing a therapist at twenty-eight years old. Her firstborn had turned one, and Alexis began to deeply feel the loss of her own mother. She fell into a heavy depression that affected her home and work life. She told her supervisor that she felt like an imposter because she was trying to hold it together for the young people she worked with while struggling to hold it together on her own. Her supervisor recommended specific professionals, and Alexis's own therapy journey began.

Since then, Alexis has worked to increase access to culturally relevant therapy for Filipinos across the state of California. She was one of the first to ring the alarm about San Francisco school data that showed disproportionate rates of suicidal ideation among Filipina teenagers. In response, she and other Filipinx social workers created a group to address mental health and wellness with Filipinx students in the district.

Alexis and her colleagues have effectively created a safety net to catch Filipino Americans in California struggling with mental health. She helped found the Filipino Mental Health Initiative, which includes a network of professionals who can provide culturally competent services to Filipino Americans across the state. This network has generated a variety of programming, from webinars and trainings to virtual counseling and online community healing circles, that are now reaching Fil-Ams throughout the United States.

Higher Ground

American individualism tells us that our wounds are our own and that we should hide them while dwelling in shame. Ableism tells us we should pull ourselves up by our bootstraps, walk off our wounds, and move on. Imee felt depressed for so long but thought it was just part of her own "melancholy" personality. Marina felt her serious mental illness was un-Filipino. Alexis thought she was being selfish by feeling hopeless and depressed. I spent most of my life believing that my pain was all my fault.

But the Filipino notion of kapwa is an antidote to American individualism, ableism, and toxic self-blame. As Leny Mendoza Strobel writes, "Our collective memory may be wounded, but from this wound also comes a collective power to heal."

Community rituals brought me close to other bodies connected by shared purpose. Once I began to understand my own body as a body connected to others, I began to believe that my pain was not just my own. Once I started to believe my pain was not just my own, I started to believe that the pain was not my fault. Once I believed the pain was not my fault, I was able to overcome shame and self-blame. Once I overcame shame and self-blame, I began to heal through collective practice, combined with intentional self-love and care. This was my bamboo framework woven by community. This was the bayanihan ritual that transported me to higher ground.

Higher ground is not necessarily a place of ease but a place of options. As Marina said to me, these options are our ancestors' wildest dreams: they are what come when you have the support to move beyond a basic state of survival. In the Bay Area activist community, I found the option of moving into a true home, a place where I could find shelter in my body as well as in my surroundings. And in this home, I found what Alexis found through ethnic studies and mental health work, what Marina found through therapy and her academic practice, and what Imee found through nature and her Kali community online. May we all find our version of bayanihan. May we all create the conditions where we no longer have to bear the burden of accumulated trauma alone.

Unbroken Water

I wish I were like a river,
So strong and mighty.
Then I shall flow
Towards freedom.

—SALIDUMMAY DKK, "DANUM"

I had once imagined the Philippines as a whole and singular body. Mother-of-pearl. Pearl of the Orient. A romanticized place with intact families and villages, where neighbors borrow cups of carabao milk rather than proverbial cups of white sugar.

I come to learn that the Philippines is a place where drought causes cracks in caked mud; the cracks look like lightning bolts and in the cracks, there are cobwebs. Even in these thirsty regions, water is dominant. Hierarchy is organized by water.

The Philippines is also a place where rivers still run fierce through broken land, though the water also bears scars. These scars become remembrance, and remembrance becomes source, and source becomes kinetic energy for change.

~

In the 1970s, a river in the Northern Philippines was ground zero in a global battle for tribal sovereignty and water rights. The Chico River—the largest tributary of the Philippines's longest river—seduced energy industrialists with the might of its current. In 2002, I went to see this river. I lived beside it for days, and it seduced and embraced me as well.

I witnessed the river's power and how its water seemed to sing of the irresistible need to flow free.

The Chico River is to the Northern Philippines what the Missouri River is to the American Midwest. It is a lifeline, an artery, a cranial nerve that connects the vital organs of regional ecosystems. Like the Missouri, the Chico provides drinking water and irrigation to millions and waters the sacred promontories, outcrops, and forests of multiple tribes.

In 1973, dictator Ferdinand Marcos targeted the Chico River for what would have become the largest hydropower project in Asia. His plans projected four "megadams" to churn electricity to Metro Manila and to address the global oil crisis, as well as what he called the country's "critical power situation." The project also addressed Marcos's need to line his pockets with World Bank funds. What it failed to address were the needs of indigenous people, whose ancestors had thrived alongside the river for thousands of years.

The Cordillera region of the Northern Philippines, where the Chico River cuts through mountains like the soul nerve of the land, is home to more than 1.3 million indigenous people who belong to different ethnolinguistic groups, including the Kalinga, Bontoc, and Kankanaey. Of these 1.3 million, hundreds of thousands rely on water drawn from the Chico River to supply communal irrigation systems for rice and vegetable farms, and to provide fresh water for drinking, ritual, and sustenance for the native forests that blanket the Kalinga and Mountain provinces.

The Chico Dam Project threatened to break this river and its sacred relationship to the Cordillera people. It would trap the Chico's currents, inundate local villages, displace more than one hundred thousand humans, and drown endemic animals and native forests.

But the Cordillera people do not easily submit to their own destruction. Hundreds organized to resist Marcos and his plan to dam their river of life. At first, protestors tried the official means of sending petitions and delegations to the presidential palace. The Marcos regime rebuffed and shamed them, calling them "sentimental" and lecturing them on the need to sacrifice for national development. He pressured the indigenous people to lay down their sovereignty in the name of a nation that had never quite included them.

And so the people turned to civil disobedience. What followed was a campaign that bears comparison to the Standing Rock mobilization against the Dakota Access Pipeline in the American Midwest. Like the fight against the "black snake" pipeline that courses beneath the Missouri River, the struggle against the Chico River Dam spanned years and magnetized the support of indigenous people and non-native allies from around the world.

In the Cordilleras, indigenous women led the way. When officials came to survey the land, Bontoc women chased them off. When armed guards came to the upriver town of Tinglayan, indigenous women performed lusay, stripping down to the waist to bare their breasts and tattooed torsos, to project bad luck on the intruders.

When workers constructed dormitories for construction camps, Kalinga women dismantled the dorms with their bare hands. When the workers rebuilt the dorms, the women returned to dismantle them again, even under threat of death. The fourth time that the dorms were rebuilt, a group of two hundred fifty women tore down the camp. Defying curfew and marching silently through the night, the women walked twenty-two miles to deliver the broken camp to police headquarters in Bolinao.

The actions of the Kalinga and Bontoc women made it clear they would go to any length to defend their river of life. But rather than back down, Marcos retaliated with soft and hard force. He bribed locals with truckloads of basketballs and chocolate and sought to "minimize opposition" by arming one tribe against another. He deliberately fomented older conflicts that had since been resolved and sowed the seeds of modern-day tribal war.

The Marcos regime also tried to break the opposition by bribing indigenous elders. They succeeded with a few, but when they lured Butbut elder Macli-ing Dulag to a fancy hotel and presented him with a thick envelope, Dulag reportedly responded, "This envelope can contain only one of two things—a letter or money. If it is a letter, I do not know how to read. And if it is money, I do not have anything to sell."

Macli-ing Dulag became a galvanizing force against the commodification of the Chico River. Dulag leveraged the traditional peace pact, or bodong, to unify Bontoc and Kalinga people against the dams. This leverage culminated in the Bodong Federation, an alliance of one hundred fifty indigenous leaders who agreed to present a united front to keep the Chico River free.

Meanwhile, the Marcos regime intensified their repression. They gave security forces free rein to execute anyone who entered so-called "free-fire zones" around the river. To defend themselves, many indigenous leaders took up arms. Blood flowed on both sides. The violence peaked on April 24, 1980, when the Philippine Army's 44th Infantry Battalion gunned down Macli-ing Dulag in front of his wife while he fixed a lock to the front door of his home. Dulag died instantly from ten bullet wounds.

Marcos hoped Dulag's assassination would demoralize the resistance, but it had the exact opposite effect: Dulag became a martyr. His death produced a surge of global protests and media coverage against the dam projects and Marcos's human rights abuses. This international condemnation made the dams untenable; Marcos and the World Bank were forced to abandon the project once and for all.

Shortly before the army murdered Macli-ing Dulag, he prophesied: "The question of the dam is more than political . . . the question is life . . . [the land and water are] sacred, nourished by our sweat. It shall become even more sacred when it is nourished by our blood."

The Cordillera people's victory against the Chico River Dam rippled across the world. As a result, the World Bank formed stricter operational guidelines for projects affecting indigenous people everywhere. The successful struggle helped produce the 1997 Philippine Indigenous Peoples' Rights Act, a law that requires informed consent from indigenous people for any project that affects their land, water, and livelihoods.

Because of the Cordillera people's organized resistance, indigenous people's rights around the globe were strengthened, and the mighty Chico River remains unbroken—for now.

~

It is April 22, 2002, and I am on my way to the Cordillera People's Festival, a five-day gathering in the mountains of the Northern Philippines. The festival commemorates the martyrdom of Macli-ing Dulag and celebrates the Cordillera people's ongoing defense of land, water, and life.

Twelve of us travel as a delegation of the Committee for Human

Rights in the Philippines. Like the twelve tributaries of the Chico River and like the twelve cranial nerves that flow from our brains through our bodies, we are wandering from our source in the California Bay to the mountains of the Cordillera Autonomous Region.

My friend BJ and I are riding a jeepney, or on top of a jeepney to be exact—top-loading, as they say in the Philippines—and the jeepney is ambling its way up serpentine mountain roads. I sit with my puwit in the hole of a spare tire, grabbing onto thin metal poles so as not to fall off the roof onto the road. We are with our friends Lisa and Drew, and the four of us have gone ahead of the rest of the delegation to help prepare our camp for the festival.

I've come on this trip to learn more about the country that my grandfather gave his life to make free. I've also come to learn about what I envisioned as paradise—a paradise of social movement resistance driven by indigenous harmony and wisdom. I'm writing an article for *Filipinas* magazine to do my part in sharing the importance of the festival with Filipino Americans back home. Along the way, I'm hoping I can distract myself from intensifying chronic pain and learn how to feel less isolated from others. Isolation is a phenomenon that doesn't seem to exist in the Philippines, in a place where people constantly gather in large groups, by choice or not, moving through their days in a sort of amoebic recombination with other human bodies.

"Are you gonna share the throne?" BJ asks, interrupting my thoughts. He shifts from sitting cross-legged to a straddle, hooking one foot under a top rail.

"Trust me, this tire is not as comfy as it looks," I reply, adjusting my bottom in the hole.

"Well, lemme try it," he insists.

The jeepney is cruising at no more than twenty miles per hour, so it seems like we can do this. We perform an awkward shuffle around supplies strapped to the roof, grasping onto the metal rails and squatting low, but even so, my vertigo gets the better of me. The greenery around me spins, I lose my footing, and suddenly I'm awash with resignation: *I am always a step away from disaster.* As I spiral with negative thoughts, my knees bang onto the jeepney's metal roof, sending a jolt of pain through my hips and back. I plop my bottom back into the tire and grab a strapping rope to steady me.

"Uh, nope," I gasp. "Jesus, you almost killed me with that idea."

BJ laughs. "We're so American we need seat belts to top-load on a jeepney."

We settle back into our perches for the remainder of the ride while I clench my teeth against shooting sacral pain. It's hotter than a pig's belly out here. I've thrown on a brightly colored batik-print cloth to shield my shoulders from the sun, and BJ has wrapped a white T-shirt around his bald head. Lisa and Drew sit on the opposite side of the jeepney roof. Between us are Kim, Randy, and Grace, members of the indigenous youth group that has been our host for the past several days.

We approach the village of Dupag, where the Cordillera Festival will be held this year, a site of one of the four thwarted Marcos-era megadams. The road narrows into one navigable lane. Ribbons of palay occupy the lane beside us, where farmers have raked the unhusked rice out on the pavement to dry. The palay stretches to the horizon, turning from fresh green to gold in the sun.

My enjoyment becomes a full-body experience when I suddenly sense a complete change in worldview. As we get closer to Dupag, I not only see the Chico River more clearly, I feel that the air is at once thicker and more clear. The trees are unbridled, bent and gnarled and wild. I can now smell some of their tangy pine odor even through the exhaust. This, I think, is a glimpse of the lushness of the Philippine Islands before deforestation. It is a glimpse into an ongoing way of life that treats nature as one with humanity.

In my nerves, I feel that the unkempt trees, the undulating rice fields, and the free-flowing river might fully envelop me. For a moment, I forget about my pain as I inhale the balm of my surroundings through my skin. And then I recoil, reflexively retreating from this full-body immersion to a familiar place of distanced analysis in my head. The worldview of oneness is seductive, but my body is more comfortable with separation.

The jeepney stops at the side of the road. BJ and I climb down. Lisa and Drew follow, and we stretch our legs and shake out our butts till our limbs go from numb to normal. Or, at least, normal for me. While the payoff for top-loading is the view, the intensified pain in my sacrum is the cost. It has now increased to a four-alarm fire. Distraction is my best medicine, so I start messing with BJ by pulling off the white T-shirt wrapped around his head. When he grabs it back, I pop an Aleve and swallow it dry, then make a mental note to find some of the weed that grows prolifically in these mountains.

The four of us stop to take a photo by a sign that reads CORDILLERA FESTIVAL with an arrow pointing down the mountainside. We are all smiles and thumbs-up as we cheese for the camera. Then we hike down a steep, rocky trail, and I gasp when I see the Chico River emerge before us.

How to describe the Chico? I have seen the Chicago River, which is efficient and architectural like a middle-aged stockbroker. I have seen the Missouri River, which is ponderous and deep, like a wise and well-seasoned elder. I have seen the Pasig River, which in parts of Manila looks more like land than water, thickened with plastic bags and Styrofoam cups and shreds of litter—immobilized like Lazarus, waiting to rise from the dead.

Meanwhile the Chico runs before us like a vibrant young being, exuberant and bursting with potential. It is a waterfall in river form. At this juncture, it stretches about a hundred feet wide, and its water is a blend of turquoise and white where rocks break the surface and turn the river into riffles.

It hits me that I am also on this trip because of water. It's been three years since I moved to the Bay Area from the East Coast, and this voyage to the Philippines is the logical outgrowth of my move toward the Pacific to be closer to ancestral shores. I feel a nascent sense of identification with the river's journey as it winds through the mountains, eventually emptying into the Pacific Ocean at a point far beyond this horizon.

I am immediately drawn to the Chico River in theory but intimidated by its physicality. It is not a river that one wades through by foot. To forge it, you must cross by boat or by bridge. Before us is a bridge erected specifically for visitors to the festival. It is an elegant feat of indigenous engineering. The bridge is made entirely of bamboo and arches several feet above the churning water, like a dolphin cresting over the river.

I step forward but then hold back, once again recoiling from nature's beckoning embrace. "After you!" I say to BJ, Lisa, and Drew, who all smirk at me as they walk ahead. In the end, I cross. Two days later, when the rest of the delegation joins us, our friend Roseli slips off the bridge and falls into the river. She scrambles back up where the bridge flattens into a walkway, but later tells us she felt as if a spirit tried to pull her in. The river seems to have arms and hands and its own will, like it is literally alive.

~

Our first evening in Dupag, BJ, Lisa, Drew and I stay with a host family as we wait for the rest of our delegation to arrive. Their home is small yet majestic, built from narra wood, a rich brown timber used to make houses and also to carve statues of rice gods.

We sleep in a sparse open room on the second floor of the host family's home. As we lay out our malongs—simple cloths sewed into tubes like thin sleeping bags—I look up to admire the thick, dark beams supporting the vaulted roof. And then I see it: a long glint of silver. It is an M16, wedged upside down in the rafters, far out of a child's reach but only an arm's length away for any adult who might need to defend against invaders. At home in the United States, I am uncomfortable with guns in the house. But here, things are different. Here, I will sleep more soundly with the M16 nested above us. It's a reminder that this village and its river are vibrantly alive only because so many have fought to keep it that way.

I begin to drift off, thinking about the Kalinga people and decolonization. According to some indigenous people, decolonization involves restoring indigenous sovereignty, preserving indigenous relationships with nature, and returning stolen land to the original residents who stewarded it for thousands of years. This would not be a regression—there is no going back to a past before capitalism and imperialism. But it may be our only way forward, especially in the context of growing climate disruption. I fall asleep and dream: What could happen if indigenous people, and all defenders of nature, led the next phase of evolution toward true interdependence?

~

In the morning, other delegations begin to arrive, and we go out to the mountainside to prepare for the festival. Each group is to build their own shelters, contribute to cooking and cleaning, and participate in workshops and performances to show solidarity with the indigenous people's movement in the region.

As the sun pulls itself up over the ridge, BJ and Lisa and Drew and I set about building a shelter before the arrival of the full heat of day. The four of us try but succeed only in being really American. City American to be specific. For some reason I'm wearing hoop earrings, and one of the earrings catches on a bamboo stick propped against a tree. I remove the inappropriate jewelry and go up the mountain with Lisa to collect stumps of wood.

We have trouble carrying them, so we roll them clumsily down the slope instead. My sacral pain throbs, but I've been too shy to ask any of the community members about weed. I'm not even sure what to call it since I only know the Tagalog slang "damo." But Tagalog is a foreign language here, so I keep popping Aleves and continue to try to be useful.

Drew is trying to split the wood that Lisa and I rolled down the mountainside, but since we are on a slope, the log just won't stay put. He swings an axe and grazes the rolling log, almost piercing his ankle instead. BJ tries next and does exactly the same thing.

Meanwhile, all around us, the mountainside is transforming into a pop-up village before our eyes. With swift bolo swipes, community leaders transform bamboo into buildings. They cut long bamboo stalks to size and bind them together with thin bamboo strips. These become the framework for dozens of makeshift dwellings as well as the large communal kitchen. Women and girls cut flat slats to make mats that serve as dwelling floors. They also craft an intricate plumbing system in which bamboo tubes act as pipes, channeling water from a nearby stream to the kitchen for cooking, drinking, and cleaning.

By afternoon, there are more than a thousand people constructing this bamboo village. Having failed in our mission to build a shelter, the four of us head to the kitchen to see if we can be useful there. For handwashing, there is a rack of bamboo with holes pierced in it. We wet our hands with the water streaming like a fountain; it feels like playing in a sprinkler. Behind us is a cauldron big enough to fit a pig, and we squeal when we find it filled with strong, black, bubbling barako coffee. We dip cups into the cauldron and sip from them all day. On a platform to the side of the kitchen, Drew and BJ begin butchering meat for the midday meal. A local shows us how to pull a bright blue bar of detergent across a pig's intestine to clean it out for pinapaitan the next day.

By evening, some of the villagers from Dupag take pity on us. They know we have no shelter—they had watched us with amusement as we tried to split wood—and so they build a wide A-frame tent from bamboo and cover it in blue tarp within the course of an hour.

~

The Cordillera people use bamboo for temporary structures because they know it is sustainable. This is just one example of how they steward indigenous forests for the sake of preserving sacred land and water. In the Ifugao region south of Kalinga, certain trees like the alimit and tuwol are never harvested because they maintain groundwater for rice terraces and forest lots used for sustainable timber. Century-old trees like the balete are also preserved because they are inhabited by anitos, or the spirits of ancestors. When they do harvest, Ifugao people choose only mature timber trees in a practice of selective cutting. And before one is cut, a local conducts rituals to ask permission from the ancestors. When they receive permission and a tree is harvested, they immediately replace it by planting two seedlings.

This system not only preserves the forests while ensuring adequate wood, it also preserves the water supply. The Cordilleran concept of inayan is key to this process. Inayan involves rituals to preserve land, water, and forest to harmonize the interdependence of human and natural resources. Kalinga culture bearer Sapi Bawer takes this interdependence a step further: he says those who live in the Cordilleras maintain the forests and the river not just because they live from them but because they are one with them—their identities are one and the same.

When we first arrived in Dupag, I sensed this unity because it permeated the air. It affected my body like a change in atmospheric pressure. In this context, my body was not just my own, and so my pain was not just my own. Pain has a purpose when our ecosystems are violated. At first, I retreated from this interdependence, just as I retreated from crossing over the river. But after a day of dwelling on the mountain, immersed in village life and enveloped by the land, my body began to dissolve its unnatural boundaries.

I go down to the river at sunset and gingerly dip my feet into the Chico's fluid at a point where it floods the pebbled bank and forms a shallow pool. I recall when I first waded my little four-year-old feet in the waves of Lake Michigan. It was a favorite moment from childhood, the first time I felt alive. The Great Lake's water was chill. The Chico's water is warm, almost body temperature. On the opposite bank of the river, I see glints from other delegates' lamps as they light their way through the forest. The sky and air are lavender, and my nerves seem to entrain to the rhythm of the Chico's flow.

~

The next morning, April 24, 2002, we reunite with the rest of our delegation, and the Cordillera Festival begins. It opens with a pattung, a sacred Igorot dance where six elders play handheld gongs called gangsas and move in a winding, serpentine pattern. The pattern reminds me of the zigzagging pain that sometimes snakes across my back. It also reminds me of the shape of the Chico River.

The barangay captain of Dupag, speaking through an interpreter, thanks us for coming to a place with no road. He says the Philippine government did not provide any infrastructure to the village, and so they had to build their own bridge.

"This is a symbol of how we build our own path," he says. "The Philippine government continues to attack the Cordillera people's livelihood by advancing mining projects that rape the land. The San Roque Dam project threatens to inundate indigenous villages, much like the Chico Dam would have done in the 1970s. But the Cordillera People create a different way."

All around us, beautiful brown humans dot the mountainside as far as we can see. The majority sit on a sloping field in front of the opening ceremony stage. Other groups are spread out, gathered under shelters, or standing to unfurl banners on higher ground. Altogether there must be more than two thousand people populating this makeshift village. The barangay captain welcomes delegations from the Bontoc, Ibaloy, Tingguian,

Ifugao, Kankanaey, and Kalinga people and announces the renewal of bodongs that will occur over the next few days. I imagine this to be a semblance of the broad unity that stopped the Chico River Dam in the 1970s, and I am thrilled to be part of it. My body relaxes into the sea of bodies that surround me. This is the paradise of social movement resistance I had come to see.

After the teach-ins, I tell one of the workshop leaders about the article I'm writing for *Filipinas* magazine. He motions me over to another tent, where he introduces me to someone named Francis. "Nice to meet you, Francis," I say. "You too," he replies in English. We make small talk for a few minutes, and then I ask what the Cordillera Festival means to him. It is only then that I learn that Francis is Macli-ing Dulag's son.

"It is important that so many continue to remember my father and what he died for," Francis says. "Especially because the bloodshed that took my father's life has not ended."

He continues, "Military and paramilitary continue to make tribal war. Tribal war is good for the corporations who want to continue mining. These threats to our people are the same as the threats to the river and to our land."

Francis explains that corporate and state interests know that if they break the Cordillera people's relationships to one another, they break the will that keeps the region's resources under indigenous control.

"How do they try to break your relationships?" I ask him.

"Just last week, two people from my village were shot," Francis replies matter-of-factly. "They were a couple on the way to their farm. Now they can no longer farm their land. And this makes other people afraid to tend their land because they are afraid they will get shot too."

I am humbled by this exchange. Clearly the Cordilleras were not the "paradise" of resistance and indigenous harmony I had imagined. I feel ashamed of my naiveté and First World motives. But I also feel a glow in the gut that radiates through my entire body. My bodily state is shifting from an intellectual understanding of this movement to a gut-driven connection with a life-or-death cause. I've found my grandfather—and his will to die for his country's freedom—in my nerves. It's no longer like the burning fire of war. It has become like the steady commitment of water.

There is a Cordillera legend that says that Kadaklan, mighty ruler of the sky, created the mountains from rocks villagers had used to stone a sorceress to death. Kadaklan created the mountains and hills, around which the rivers must now wind, to remind the local people of the futility of bloodshed. But Kadaklan's lesson must have been aspirational. Perhaps Kadaklan did not yet know of militarized dams and mines and public-private partnerships for resource extraction. When so many forces come for you, it is hard to imagine how a warrior people at one with their ancestral domain would simply lay down arms and surrender.

~

In one of the most common photos of Macli-ing Dulag, his eyes are intensely cast at an unseen horizon outside the picture's frame. His brow is furrowed, and he wears a look of determination, a look that is at once strong and gentle. The bags under his eyes recall the cascading rice terraces of the region, and a line runs from his left eye to his cheekbone—whether it is a scar or a scratch on the photo I do not know, but the effect is like the track of a tear.

Today, Dulag's determination and his tears are carried by the next generations. More than forty years after Dulag gave his life to keep the Chico River free, the Chico River continues to attract authoritarian ambitions. When Duterte was president, he tried to dam the Chico under his national program to "build, build, build." Now that Bongbong Marcos, the only son of former dictator Ferdinand Marcos, is president, the fate of the Chico—and its indigenous defenders—remains to be seen.

History repeats itself unless we break harmful cycles. History repeats itself when we lack a revolution of values that protects bodies of water (people, rivers) and land (nations, earth). The old colonial strategy of divide and conquer is at work here. Developers broker deals by promising cheaper electricity and consistent irrigation to upland communities at the expense of those downstream. The indigenous people of the Cordilleras are familiar with the game of divide and conquer. They rebuffed Spanish colonizers and to some extent resisted American rule. But if history repeats itself in good ways more so than bad, then perhaps the Cordillera people's self-determination will prevail.

For me, the gift of attending the Cordillera People's Festival was a new way of seeing and also a new way of being in my body. It was the gift of going beyond divides of self and others, humans and nature, illness and health. I saw the Chico River, touched its waters, and knew this is how I wanted my body to be: undammed and life-giving, in symbiotic relationship with its natural environment. It is what I believe all our bodies deserve. To achieve this, we must be in symbiotic relationship with our social and political environments as well. Just as we need a Civil Rights–inspired revolution of values, we also need an indigenous-inspired commitment to reshaping ecopolitics over generations.

During the Cordillera Festival in 2002, I also interviewed Banag Sinumlag, a Kalinga leader who was involved in the Chico Dam struggle in the seventies. When I interviewed him, he was sixty-seven years old and a full-time organizer. Twenty years later, while writing this book, I encountered a news article that featured him and learned that Sinumlag is alive and continues to fight against the new proposed dams. He is now in his late eighties, and despite threats of violence, he still farms. But he takes days off to steward the water, and his grandchildren have joined him in the ritual.

~

The story of the Chico River is an unfinished one. It is an allegory for the place where all of humanity finds itself now: at a crossroads between continued exploitation of natural resources and a more sustainable way of being, which is essentially the crossroads between climate destruction and transformation. It's hard to see how we might choose the right path without completely changing our worldview.

Nature functions as an autonomic nervous system regulator, perhaps because we can entrain to external rhythms that our bodies find soothing, including the flow of river water and the blowing breeze. How much more could we increase our collective well-being if we embraced nature to the point of defending it like we would our own bodies?

I lived on the banks of the Chico River for five days in deliberate community with thousands of people whose ancestors have for centuries called that watered land their home. The twelve of us delegates washed clothes and drank from the river's waters. We danced and sang and learned about organized resistance in the river's defense. The sharp edges of my pain smoothed to a dull and bearable ache, and I realized that normally, in my regular city life, I chronically braced my body, as if I needed to maintain a dam between me and the rest of the world. Here, I became so fully embraced by nature that there was no room for the myth of separation.

At one point while I sat on the mountainside looking down at the river, I felt a tap on my shoulder. I turned around, but there was only a tree. Its branch must have brushed my back. And then I heard a voice in my head, and I somehow knew it was the voice of Bathala, the supreme goddess in Tagalog tradition. *I'm here*, she told me, *on this mountain, in your body, in the river and the trees. I'm here, and I have been here all along.*

~

It is the final evening of the Cordillera Festival. The teach-ins are done, the cultural presentations shared, and most of the food has been eaten. Bodongs have been renewed. Many delegations are packing up camp and clearing the mountainside of litter while others dance and drink rice wine.

BJ and I go down to the riverbank with a few other delegates to watch the sun drown in the water, as Tagalog legend describes the sunset that leads to the next world. The Chico gurgles. Suddenly, we see two strange figures fly up from the water and seemingly disappear in midair.

"Did you see that?" BJ asks me.

"Yeah. What was it?" I reply, feeling confused.

"Not birds."

"No, they were, like, rectangle-shaped. They looked like, flying folders or something." My arm hairs are standing on end. There is a thickness to the air that tells me the anitos are congregated. I feel at peace, even accompanied by them. The vision we have seen is to me just a part of their inexplicable presence.

"Let's go back up," BJ says.

"What, are you scared? Of folders?" I elbow him.

"Those weren't folders. Bye," he says with his back to me. "I'm not trying to mess with this river."

He scrambles back up the gravel path to the festival site, and I follow him, but not before looking back to see the river transform into a fire opal as the sun falls away.

Back at the festival site, we're greeted by the thick smell of weed. Our delegation is passing a fat joint, courtesy of one of our new friends. "Yessss!" I cry out with joy, running past BJ to get to the joint before him. One of our delegates talks to the person in charge of the sound system, and the next thing we know, the loudspeakers blare Bob Marley's "War," and everyone on the mountainside is dancing. The indigenous youth delegation brings coconut shells filled with basi, but before they hand them over, they point at the joint. We laugh and gladly make the exchange. After a shell full of rice wine and several tokes of strong weed, I tell BJ I have an idea.

"What?" he says, too busy dancing to even look up.

"So you're afraid of the river, huh?"

The next morning before we leave Dupag, BJ and I climb to the top of a cliff along with another one of the twelve Filipino American delegates. The three of us squat at the edge of the rock, peering down at the mighty river below. Everything around us is relentlessly alive. There are dozens of people in as many wooden boats dotting the river from bend to bend. Hundreds of people on the Chico's banks squat to wash clothes, smoke cigarettes, or carry pigs strapped to poles. The river itself churns, and I imagine the spirit that tugged Roseli stirring its very depths.

We scamper up the cliff. I remove my glasses and become near blind as a bat. I can't see, but I can feel the surrounding web that will catch me as I fall. The people, the river, the anitos. The men in their boats laugh and cheer the Americans to jump.

"You first," I say to BJ.

"What?! This was your idea."

"Come on, it's your initiation. You're the scared—"

Before I can finish my sentence, BJ jumps and everyone below lets out whoops of jubilation. Our other friend goes next, and the whoops arise again.

That leaves me. I kick off my tsinelas and watch as gravity makes their fuzzy outlines creep toward the cliff's edge. I hesitate. I have a flashback to when I nearly jumped off a Fort Funston cliff, far away from where water could wash me away. My body is still in pain, but that pain is no longer contained only in my skin. It is dispersed, borne by the river and the mountains. The scars and the bruises of the water and the land are one and the same with my pain. I look down at the river and I see myself. I leap.

The Chico is cold as it swallows me. Its throat is thicker than water, thick almost like blood. And then it releases. I breach its mouth, and the river once again shows its limbs, which are the crests of its currents, and they hold me like a parent rocking a wanted child. I float, I rock, I sing that we are from water, so may we become water, with its power to defy the physics of becoming broken.

Awaken the Lyrics

Track 1: Rise Up

Picture this:
Everybody holding hands on the lawn of the city hall
Now it's on

Working-class deep
Third World countries
And the people in the streets of San Francisco
And the whole damn Bay

We're ready!

And this is what we're gonna do:
Organize the veterans down to the youth

It's time to build for now and generations to come
Cuz we're all God's people under the sun

—DISKARTE NAMIN

In the spring of 2000, the Bay Area bustled with a rising culture of youth organizing. The 1990s myth of the "superpredator" had led to an epidemic of policies criminalizing young people of color. In response, youth leaders grew a movement to push back against gang injunctions, police in schools, and Proposition 21, a proposed California state law to try kids as young as fourteen as adults. Meanwhile the dot-com era caused rents to skyrocket in San Francisco. Speculators razed old buildings to build expensive live-work lofts. Evictions soared in poor and working-class communities like the Mission, South of Market, and the Fillmore. Young activists joined with experienced organizers to fight for stronger tenants' rights and more affordable housing to keep immigrants and families from getting pushed out of the city.

Along with this wave of organizing came a rising tide of underground music. The sound of the Bay in the early 2000s was the hyphy movement slap of E-40 and Mac Dre, the political hip-hop of Zion I and The Coup, the Afro-Latin conga-driven dance songs of O-Maya and Agua Libre, and the heavy tuba womp of rancheras blasting from car stereos in the Mission.

I arrived in the Bay during this upswell and was electrified by what I heard in the streets. While working as a fellow at a national magazine, I interviewed a local DJ, a member of a crew called the Local 1200, to get his take on Prop 21. I was fangirling and so intimidated that my voice shook terribly when I asked him questions. At that time, I couldn't have dreamed that I would one day find my own steady voice by becoming part of this political music scene myself.

For much of my life, speaking out loud landed me in desolate terrain. I had a sense, even when I was small, that words were dangerous. Instead of talking, I found solace in music. I took shelter under my father's baby grand, and, in the basement of our red brick house on a hill, spent sunny afternoons with an 8-track player. As my brothers played eight ball, I popped chunky plastic bricks into the 8-track's rectangular slot. Out came the deep bass of "Another One Bites the Dust." *Da dum doom doom doom,* and I bopped my little head to the beat as those rumbling notes fell over and over and over again.

Listening to music brought a warm glow to my chest. Armor fell away from my heart, and energy seemed to burst from my lungs. This feeling was especially strong whenever I listened to live music. It was never more powerful than one night in the Bay in March of 2000, when I first heard the band Diskarte Namin rock a full set.

It was an unusually hot night at the Cell, an old screen-printing warehouse turned crucible of community art and organizing in the east Mission of San Francisco. At the Cell, youth groups like Loco Bloco practiced drumming and Sisters of the Underground taught b-girl classes. Organizing groups like United Playaz, Olin, and Homey held workshops and strategy sessions. The Cell was just one of the many spaces where I would grow to feel at home as part of an activist family.

That night, I was with two of my newest friends, BJ and Mia. When Diskarte took to the stage, the lights dimmed and a spotlight switched on, bathing the band in blue. J, a tall kalbo dude who called himself "The Brown Commander" ruled the throne behind the drums. Paul, a guy with a goatee who was often mistaken for Mexican and whom J called the "Pizzaslayer" rocked out on both lead and rhythm guitar. J's stocky cousin the Ricker swayed back and forth on bass. And Big M took center stage with the mic. They came out the gates running, opening with what would become their signature song for the next few years:

Rise Up! Rise Up!
Rise Up! Rise Up!
Reclaim the streets put your fists in the air
If you're down with this put your fists in the air

Big M called and we responded, the whole crowd throwing our fists up as one. The way Big M performed was from the guts. True to his name, he was a big Filipino dude who used to joke about being a small Samoan. As he rapped, he dripped sweat, and he wiped that pawis off with a towel slung over his Warriors jersey. His whole body was in the music, and it was his movement as much as his voice that pulled me in—we could feel the rhythm of his rap like the rhythm of a sledgehammer driving through a concrete wall.

I jumped like a rabbit and raised both arms in the air, reaching, reaching. Unlike the staid classical music my father played around my childhood home, Diskarte's music moved me with its raw power and unpredictability. Each hit of the bass drum made my heart pound.

When I was a child, my father wanted a family string quartet, and so I started Suzuki violin lessons at age three. I strained my little neck to clamp the violin between chin and shoulder, which I didn't like because it hurt. "You'll get used to it," my dad said. I believed him. When it continued to bother me, I told my mother I didn't want to play anymore, and she replied with what would become one of her standard refrains: "Well, you can't always do what makes you happy." I must have been four or five, and this was the first time in life that I remember feeling angry. Why couldn't I do what made me happy? Or why couldn't I at least stop doing something that hurt? I pressed my lips together and set my jaw but remained silent, since using my words had led to no good.

Once, when I was about nine, I had to pee really badly during a violin lesson. I was practically dancing while I dug my bow harder into the strings and clenched my bladder to keep the river of urine at bay. I could have asked my teacher for a bathroom break, but I felt stuck: like a boulder had blocked my vocal cords and the option to speak had been cut off. So I stood there crunching my neck, squeezing my pelvic floor, and moving my feet in a frenetic pattern. I think my teacher just thought I was really feeling the music. And when the rat-a-tat began and my teacher asked, "What's that?" I pretended I didn't know it was the sound of my pee dripping onto his rug.

For much of my life, my body was in a state of both pain and anhedonia, the inability to move toward pleasure due to depression or trauma. For people with anhedonia, neural pathways involved with emotion become downregulated as part of an adaptive trauma response. I stopped feeling because there was only so much pain from isolation that I could take.

Diskarte's music awakened me from this numbness, and it felt electric, like a rush of adrenaline mixed with an overdose of joy.

The Cell was packed that night. The audience moved like one tentacled creature, bathed in that ice-blue spotlight that looked cool but made us all sweat even more. The music and motion permeated my skin and traveled through my nerves. Listening to "Rise Up" and chanting along made me feel capable of explosive and dazzling things.

I wasn't a rapper, I wasn't even a hip-hop head, I was just a Filipino American girl from the Midwest raised by music. I also had an instinct for what was good for me. And that night at the Cell, experiencing Diskarte Namin perform, I knew I wanted to be a part of this band.

While we watched Big M rap, BJ yelled to me above the music, *I wanna join Diskarte!* I yelled back, *Me too!* and gave him a pound. *Really?*

Let's do it! What would you do? he asked, still bobbing his head to the music. *I wanna sing!* I exclaimed, still jumping along to the beat. *Me too!* he yelled back, and this time we gave each other a high-five.

It turned out that J was recruiting. J was a community organizer like his then wife, Rachel, and he was shaping Diskarte into a cultural arm of the Committee for Human Rights in the Philippines. This meant he wasn't just recruiting performers; he was looking for cultural workers and protest musicians.

BJ fit that bill to a tee. With years of cultural activism under his belt as a Filipino theater performer and Afro-Caribbean dancer at Sacramento State, he was a shoo-in for the band. Meanwhile, I sang English art songs in high school choir and college opera and a cappella. I shouldn't have made the cut. But I belted "Real Love" at an audition at J's house, so he knew I could sing, and I met Rachel at the WTO protest in Seattle, so she knew my politics were aligned.

J welcomed me into the band, and for the next ten years, even when I reverted back to old habits of avoidance where I tried to disappear, he always reached out to bring me back.

Diskarte Namin means "our strategy" in Tagalog, an apt name because we used music as a strategy to mobilize people to action. Joining Diskarte was also my own strategy to navigate an exciting but uncertain new world of Filipino American activism while staying grounded in a musical language that had accompanied me since childhood.

BJ and I started out as roadies and the sound crew. I didn't mind. I was used to being behind the scenes. Besides, I was thrilled just to be invited in. BJ lugged amps and monitors while I set up mic stands, adjusted booms, and connected audio cables. The amps were often sticky with sweat and spilled beer. The mics sometimes smelled like bad breath. I quickly learned that this part of being a musician was not about romance; it was about ritual. Coiling cables at the end of each gig, applying that gentle twisting motion to allow them to loop at their natural bends, is just one example of a roadie ritual that became soothing to me. To this day, when I see audio equipment—the black boxes of amps and speakers and monitors and mixers and the metal of XLR and quarter-inch cables—I feel fully in my comfort zone, and I get the urge to jump on stage and sing.

For a while, BJ and I would set up and sit on the sidelines, monitoring sound. Then BJ started bringing his kulintang, a set of indigenous gongs from the Southern Philippines, and J gave me a cowbell and a cabasa to play, as well as a mic to try out backing vocals for Big M's raps. My first show onstage was for a Philippine human rights conference at East LA Community College in the fall of 2000. My debut involved screaming and banging on the cowbell as we performed our closing song "Kalayaan"—a rap-metal rager that left our lungs out on the stage.

The sea of audience members threw their fists in the air as we wailed, *Kalayaan para sa bayan!* And Big M began to chant, *Ang tao, ang bayan, ngayon ay lumalaban!* I felt elated by the way the steady pulse of the music, combined with the chant, brought hundreds of people together. We were screaming and chanting and pulsing our way into something new, into a collective culture of dissent. Musician Christopher Small has observed that when you take part in a musical performance, you are actually saying to those around you, "This is who we are." This is who we were, part of a rising culture of West Coast activism that built relationships of trust that continue to this day.

The process was both psychologically and physically healing. Screaming was cathartic. I had never screamed so hard in my entire life. It was like breaking through walls of generational silence. And at that time, my chronic pain was intensifying, to the point where I periodically lost the use of my arms for hours on end. Yet I could still bang on a cowbell and scream like the stage was on fire, like my life depended on it. And maybe it did.

I now know that music is an attention modulator. When it captures your focus, it helps distract from negative sensations and emotions. Performing with Diskarte was an analgesic that made my dopamine and endorphin levels rise up. Maybe this is what Bob Marley meant when he asked to be hit with music, because when it hits, you feel no pain.

Track 2: Diaspora

There are times when I am one with the field
And there are times when the sun soothes my soul
In the mornings, dusty winds take my pain
Then in the evenings, an old memory fades away

After that show in East LA, I became an official member of the band. Eventually, I earned status as co-lead singer along with BJ. We had graduated from roadies to front people, and I was loving the spotlight, even though I still struggled with anxiety and pain.

Spending time with Diskarte reminded me of my best moments growing up, listening to music and goofing around with my brothers. The guys in Diskarte became like siblings—people to laugh with, but also people with whom I could find release for the martial energy in my nerves. When we weren't channeling this militance onstage, we hung out with a lightweight rhythm, floating from topic to topic and noodling around with songs. This easy camaraderie felt safe. We spent rehearsal time drinking Jack Daniel's and clowning each other, and then we'd go out for tortas at 2 a.m. and joke around some more.

We all had day jobs but gathered about three times a week in the evenings for practices and gigs. For the first couple years, we didn't get paid—being asked to play felt like compensation in itself. There was something special about having a calling to devote ourselves to beyond the daily grind of bills and building a career.

In early 2001, at a hole-in-the-wall rehearsal spot called Time and Space, Paul showed us a song he was working on, a very different protest song from "Rise Up," a ballad called "Diaspora." I remember sitting on the corner of the stage, a little too close to a closet with a toilet that was just a sludge-filled porcelain bowl with no plumbing. I caught whiffs of that odor mixed with cigarette smoke from the alley while Paul sang a couple verses he had drafted over a beautiful guitar lick. "Diaspora" told the story of a Filipina mother who left home to make money overseas.

I didn't know much about Filipino migrant workers at the time. But the band studied together. We learned about Philippine history and contemporary politics. We learned that the Philippines's largest export is human beings, and that some of these human beings are women who end up trafficked into the sex trade.

This was a far cry from the English art songs and Italian arias I had memorized in college. It was even a world away from the whimsical lyrics of Queen and Madonna songs I parroted as a child. "Diaspora" was an original song about a contemporary problem that hit close to home. Paul's lyrics told the story of a mother who wanted nothing more than to be home with her children. Yet she had no way to feed them in the Philippines. She had to leave her country and become a migrant worker so she could send money back to support her family.

Now in the darkness
Neons flash the music starts
I dance to rhythms of a thousand lonely hearts

"Diaspora" moved through a spare conga beat, expressive acoustic guitar, and long melodic vocal lines. The song brought my body, which had been stuck in defensive silence from traumas past, into a state of hyperpresence. I had to be fully in the moment to play off the energy of the audience and to watch Paul's body language as we felt out changes in volume and tempo, making micro adjustments to keep the rhythmic guitar at pace with the elastic melody.

With presence came an ability to surrender to the lyrics, to become the migrant woman forced to leave home to provide for her children. And singing as her, singing for her, allowed me to fully embrace the necessity of forming words into stories and stories into song.

Your empty promise to provide
But there is no way here
We can't survive
And so I spread my wings and fly
From home

Whenever I sang "Diaspora," I thought my heart would fall out of my mouth. I felt the collective grief of Filipino overseas workers in my chest. As I sang, this grief moved from my gut through the microphone, into the venue, and through the audience. I saw people in the crowd look alternately angry and bereaved. It was a shared mourning in musical form.

I've since learned that music can change the brains of both performers and audience members to create a shared state of entranced presence. This entranced presence is conducive to collective feeling and, in the context of organizing, collective action. The role of a culture worker, I realized through songs like "Diaspora," is to take center stage and declare with full embodiment and soul a vision for the humanity of those whom unjust society would bury in its dust. I had always known that words could be dangerous. Now I saw them not as vulnerabilities but as weapons. Words whet with truth can expose injustice, and also glint like a beacon toward what could be.

But if you promise to provide
Food for my family and ways for us to survive
Well then I'll spread my wings and fly
Back home

Track 3: 100 Years

No more of the silence
No more will we hide from our past
We will fight for a future
Where we take back our pride in our land

As I continued to grow with Diskarte, I began to realize that my musical life before the band was simply a coping mechanism—an opiate to my pain. The real healing began when I took advantage of the unique opportunity that singers have over instrumentalists: the ability to write music through words.

After singing "Diaspora," I started to believe I could create my own lyrics. The result was "100 Years." If "Rise Up" was a collective expression of "This is who we are," and "Diaspora" was a story of what could be, "100 Years" was me standing up and saying, "This is who I am now."

In the winter of 2001, I drafted lyrics at Time and Space by the glow of the purple string lights, on the farthest corner of the stage away from the dreaded toilet of sludge. This was the post-9/11 era of the War in Afghanistan, just before the forgotten war on the Abu Sayyaf in the Southern Philippines, a time when people with Muslim names—including Muslim Filipinos—were being profiled by Homeland Security and visited by the FBI.

At that time, I felt especially agitated about the invisibility of Philippine-US relations, and so I began documenting what I knew best, which was history. I jotted notes about the Treaty of Paris, through which the United States bought the Philippines from Spain for twenty million dollars. I wrote down statistics about the number of Filipino immigrants to the United States, anecdotes about veterans denied benefits after World War II, and other key moments after the official end of the Philippine-American war. In the end, I had composed a history essay, but I couldn't make the words come alive.

Judith Kitchen wrote that in order for words to become lyric, there must be a lyre. A few months after I started writing "100 Years," J introduced us to a rolling bass line that became the lyre I needed to turn my history notes into song. At a cabin in Santa Cruz owned by a family friend of Paul's, amidst tall windows that framed towering redwoods, I molded the words into lyrics that moved with the drive of J's bass.

I don't remember the first time we performed "100 Years," but I remember what it felt like to sing the countless times we put it into a set: it was like coming out of hiding.

100 years after bridges were built
From America to overseas
Now we're two million strong in a land that is rich
With a few opportunities
No more little brown brothers sisters
No more of this bullshit about backwards friends
Now we're two million strong in America means
That we all are our homeland's revenge

After decades of reining in my voice, I had finally broken silence and unleashed my own words. The pain-relieving and identity-building impacts of performing with Diskarte were prerequisites to arriving at this threshold. No longer fully contained by my pain nor isolated from collective agency, I awakened the power of my own perspective through lyrics.

Filipina Australian writer Merlinda Bobis calls this process of breaking silence "opening the throat." Bobis considers opening the throat an act of decolonization, a rebellion in words, the generative ritual of "storying back" against colonial erasure.

Storying back is something I had once dabbled in and was chastised for. Prior to this, I had written carefully reported journalistic articles, school critiques of English literature, and personal stories in Asian American history class. When a white graduate TA almost failed me for writing a "disturbing amount of anecdotes" in an exam, and for drawing more from my own experience than from official texts, I felt the dam in my throat cemented by his words. To me, he was saying: *Your life doesn't count, even when the subject is your own history.*

The power of this white TA's words dissolved whenever I performed "100 Years." It was like serving up a forbidden lecture to younger versions of myself. The lyrics came alive to the beat of the congas, the rhythm of the flamenco guitar, and the drive of one of Big M's best raps, performed over an up-tempo version of the bass line to "Paid in Full." As I sang the soaring chorus, I would lift my palms to the audience, then bring my hands in toward my chest, inviting them to join me in the ritual of opening our collective throats.

About a year after I wrote the song, Myke, a fellow musician from DKK, a cultural arm of the Cordillera People's Alliance in the Northern Philippines, translated the lyrics into Tagalog.

Huwag nang matahimik
Huwag na ring makubil sa dilim
Bukas ay panghawakan
Itanghal sinilangang bayan

The Tagalog lyrics felt powerful in my mouth, even as I was still learning the nuances of the translation. The sound of deep Tagalog transformed the entire piece. In English, "100 Years" had a light and danceable feel. In Tagalog, the guys instrumentalized the piece with a richer, deeper, dub reggae tone. We invited friends, Ron Quesada and Ryan Leaño, to play the kulintang. The melodic gongs mixed with Tagalog were mesmerizing.

Singing "100 Years" in Tagalog elevated the song to its higher purpose: to claim a history of dignity and resistance in the face of colonial erasure. As Paul wrote in a book about cultural activism, singing in Tagalog "is an overt act of cultural nationalism necessary to avow the strength and beauty of Filipino languages and dialects from which many Filipino Americans have been disconnected." I felt proud to reconnect with my mother tongue, and to use native words to share a people's version of history through song.

Looking back, was "100 Years" Diskarte's best song? Definitely not. Do I cringe a little when I hear my voice on the recordings? Sometimes I cringe a lot. But I put myself out there and unlearned years of being told that my own ideas did not matter. I survived the fear of ridicule and wrote for anyone hungry for suppressed history, for anyone unsure that they themselves had anything to say. In other words, "100 Years" cracked open the floodgates to the writing I do today.

Track 4: The River Song

Do you know how freedom sounds?
I said do you know how freedom sounds?
It sounds like spirits in the native trees
Sounds like thunder from the ground

In spring 2002, BJ and I went on a life-changing trip with a delegation from the Committee for Human Rights in the Philippines. It was there that we wrote Diskarte's most remembered anthem: "Chico Redemption," popularly known as "The River Song."

I took a minidisc player with Paul's sketch of a new song, just a few chords: A minor, G, F, strummed with some exploratory noodling from C to E. On a ferry from Batangas to Mindoro, I hummed to the chords while a child dropped a foam bowl of instant noodles into the water. That sight led me to pen lyrics about what lies at the bottom of the bay. A week later, we attended an indigenous people's festival in the Cordillera mountains of Northern Luzon, and the song took on a whole new life. There, on a mountainside above the mighty Chico River, the lyrics to "The River Song" wrote themselves.

One evening during the festival, BJ and I crouched on the banks of the Chico, freestyling over the minidisc chords and the river's roar. At first we goofed off, talking about outhouses and kilikilis and hairy pigs turned into lechon. Then our words began to morph into the scene before us—trees, thunder, spirits. The lyrics seemed to flow directly from the river's depths. The next day, verses took shape when we heard the indigenous culture group DKK perform a song called "Danum," which spoke of the river's might, like the people's movement to be free.

While BJ and I were in the Philippines, Paul continued to develop the chord structure for the song. This was typical of how Diskarte created our music. An instrumentalist would start with a beat and a melody or a set of chords, and then BJ and Big M and I would work on lyrics to lay on top. Or it would start the other way around, like it did with "100 Years" and "Rise Up." One of us would bring lyrics or a rap, and the instrumentalists would experiment with bass lines, guitar riffs, and drumbeats to carry the words along. Then, as a full band, we would flesh out the piece in practice, doing run-throughs with different structures, adding chord changes, new bridges and solos as the music moved us while we played.

With "The River Song," the process was no different, except that there was magic sprinkled in. When BJ and I returned to the Bay with lyrics fresh from the Chico struggle, the bones of the song came together in one take. I remember Paul fingerpicking his guitar, then Mr. T beating together his drumsticks with a one, two, three, and J dropping a head-bobbing bass line. I glided in with lyrics, and by the end of the first verse, I had goose bumps. The song was flowing like the river itself.

But something was still missing. So Paul went back to work and came up with a soaring bridge that ramped the song up to a climax and breakdown. Our friend Irene Duller, formerly of 8th Wonder, wrote a passionate spoken-word piece for that breakdown, and we knew the song was done. We put it into our set for an upcoming event called the Link Arms, Raise Fists Conference, a production of Filipinos for Global Justice, Not War.

The cultural night of the Link Arms Conference took place in the cafeteria at City College, a rectangular open space with a bright linoleum floor. The room was packed with people, and I remember trembling a bit when I announced a brand-new song about the Chico River. I described it as a gift from the indigenous people's movement in the Northern Philippines.

Paul and J started playing the opening, that simple bass line that kept pulse on the downbeat and a guitar line that sounded like a river at work. Then BJ and I took to the vocals:

There's a struggle in those hills
You can feel its power still
The women take down private property
In a peaceful act of the people's will

BJ and I wrote the verses as a form of "talk story," a Filipino oral tradition of sharing heritage, news, and tsismis. While "100 Years" had been my own perspective on a key moment in Philippine history, these words were collective lyrics about indigenous Igorot history—neither the words nor the history belonged to us, they were much larger than us, and BJ and I served as the channels.

When we reached the chorus, BJ branched out into stunning harmonies that gave me chills. I later learned these musical chills happened because his voice literally changed the blood flow in my brain. Looking out into the audience during the chorus, I think others felt these chills too. I could tell from their raised eyebrows, clasped hands, and the way some people closed their eyes to savor the sensation.

A friend who used to sing with The Wailers once told me that the foundation of a good song is a melody as simple as a lullaby, with lyrics as easy to repeat as a nursery rhyme. Sure enough, by the third chorus, I saw audience members mouthing the words. And by the time Irene delivered her poem in the middle of the song, I could see folks sway as one to that rhythm. They rocked back and forth, and I saw their jewelry, the bands on their construction shirts, and the reflectors on their shoes catch glints of overhead light until the moving crowd looked like one flickering flame. Just then, a few people put lighters in the air. This was a joke, and I smiled as I sang, my smile growing wider and my eyes brimming as I saw more lighters rise. The bodies of flickering flame transformed into a constellation of stars.

I now know that this was the biological phenomenon of entrainment in action. Individual internal rhythms—blood pressure, heart rate, breathing—were all changing to align with the beat of the river. These bodily changes affect our moods. From the beginning of the song to its close, the mood in the room alchemized from interested agitation to entranced elevation.

And did you know that freedom looks like a river
Flowing free and strong like how we found each other
And yes together we will work
Try hard to make life better
The way that water carves through stone
But we can't do this alone

Toward the end of "River," which climaxes in an outro of repeated harmonies, I saw a toddler break from the crowd and waddle toward us, stopping just a foot away from the band. They swayed their diapered bottom to the rhythm before falling on their bum. People who had been standing crouched down, people who had been sitting stood up, and the effect was like a wave of bodies moving to the ebb and flow of the Chico River itself.

Afterward, people came to greet us from the crowd. They said, "Why you gotta make us all cry?" Then enveloped us in bear hugs.

When BJ and I wrote the lyrics to "River," we didn't expect the song to make such an immediate and powerful connection. But then again, we weren't that surprised when it happened. Oliver Sacks writes, "Rhythm binds together the individual nervous systems of a human community." The movement wrote the music, and the performance made its binding power come alive.

Part of this bonding effect may also come from the fact that listening to or playing music awakens the reward centers in our brains. Music increases blood flow to the same parts of our brains that activate when we eat chocolate or do drugs or have sex. Meanwhile, music turns the volume down in areas associated with depression, anxiety, and PTSD.

This explains why music has always felt like church to me, and perhaps also why music has been used by workers, prisoners, enslaved people, and others in oppressive situations throughout history—it can literally bring our bodies to a higher plane of connection and pleasure even when the conditions we face are bleak.

In this way, music can be powerful medicine—so powerful that neuroscientist Bruce Perry says that it is a greatly undervalued treatment for trauma. I believe that rhythm and melody are some of the forces that kept my ancestors alive. Music continues to heal their traumas that I carry in my body today.

But I'm sad to say music wasn't healing for all of us. Over the years, two of Diskarte's members struggled with addiction. One was homeless before he went to rehab and entered recovery. The other dropped out of the community. And at twenty-nine years old, BJ died suddenly of what seemed to be mysterious causes.

I think of all the history we carry in our bodies, of my privileged access to health care and the barriers to wellness my bandmates may have faced as working-class brown men, burdened by traumas past and generations of toxic masculinity, amid the pressures of living in a gentrifying city. It reminds me that music alone cannot heal all wounds.

But music did bind Diskarte together like family for almost a decade, and it allowed us to contribute to a culture of youthful rebellion that has since grown into multiple movements for justice. In the process, our nerves, our bodies, our baselines were forever changed.

~

We can inherit markers of trauma, but we can also inherit markers of strength. From precolonial gangsa and kulintang to karaoke and "music minus one," music courses through Filipinos' blood. This is an inheritance that I value every day, and one of the things I am most grateful to my father for directly passing down to me and my brothers.

Diskarte had the honor of being part of this musical lineage, while also cocreating a new tradition of musical resistance. In our ten years together, more than a dozen cultural activists moved through the band, including rapper Saico from the Kasamas; my dear friend Mia, who played the kulintang; and my future partner, Juan, who composed songs and played bass and guitar. We helped build bridges from the Bay to the Philippines, rocking shows with Xicano groups like Aztlan Underground and Quetzal, and Native bands like the Diné punk trio Blackfire.

"The River Song" was a watershed in this journey. The lyrics that flowed from the Cordilleras to the US flowed back to the Philippines, as indigenous artists asked us to use the song in their performances and protests. Myke from DKK, who translated "100 Years," shared "The River Song" in tribute to Pepe Manegdeg and Albert Terredano, two human rights activists murdered by paramilitary.

The song also became a vehicle for political education among Filipino communities in North America. A group of Pinay activists in British Columbia adapted the song for their choir. University of Hawaii scholar Daya Mortel wrote a paper that called "The River Song" a lasting remembrance of resistance and revolutionary love. Back in the Bay Area, J's daughter A has remixed her own soulful version, and her friend S is choreographing an original dance to the song. And at a cooperative summer camp called Sama Sama, teachers and parents developed curriculum to teach six- to fourteen-year-old kids the history of the Chico Dam victory, as well as how to sing and play "The River Song" themselves.

More than twenty years later, "River" has become a true folk anthem, taking on a life all its own. This was always Diskarte's best dream: to cast a stone that made wide ripples, to awaken a diaspora's lyrics through song.

IV
Neuroplasticity

- The capacity of neurons and neural networks to change, particularly in response to repeated use, environmental stimuli, and new information.
- Where the journey continues, tracing the possibility of reshaping our nervous systems toward interdependence.

Underwater, she endures.

—ANGELA PEÑAREDONDO

Unconditional

At first, you were just another brown and down body, one more Fil-Am activist to prove myself to. We met at a movie screening for the Committee for Human Rights in the Philippines, amid the velvet curtains and worn carpet of the Victoria Theatre. The default event attire was of the jeans-and-political-T-shirt sort, yet there you were in a pressed suit the color of ash, looking like you were ready to sell me something.

This was January 2000. A friend introduced us, and it was immediately a melding of minds. We said, "Nice to meet you" in one breath and started talking shit to each other in the next.

"Laundry day, huh?" I pointed to your suit.

"Time for you to do your laundry day, huh?" You pointed at my T-shirt with its faded "Serve the people" slogan and my ratty-looking scissors-cut jeans.

I smiled but persisted. "Is that your dad's suit?"

"What? No, it's mine. I just came from a job interview."

"Oh okay. It's yours. So . . ." I swatted at the suit's oversize shoulder pad. "Why do you look like you just stepped out of the eighties?"

"You're wrong," you replied, but you were laughing. "And hey, don't knock the eighties."

We spent the rest of the night standing outside the curtained doorway, cracking inappropriate jokes about each other during what was otherwise a deadly serious film.

That was how you were, I would come to learn. You could laugh your way out of a dark corner. But you were also a true Scorpio; sometimes your dark corners engulfed everyone around you. I think this is why, for the most part, you kept your joy turned on and your stinger tucked away.

Below your bald head and above your compact dancer's physique, your wide, dimpled smile shone like a beacon.

From this point forward, we were Bonnie and Clyde, Thelma and Louise, Han Solo and Chewbacca (I was Han, obviously). We joined a band together, we moved in together, we became a pair that marched together, smoked together, sang and danced and karaoked together.

My friend Emily later said, "BJ would give his eyes for you." And I replied, "I would give mine first."

~

I was raised to trust no one. To believe that safety can only lie in power, achievement, and material gain. I grew to believe that love is distanced and clinical, conditional and transactional, dependent only on what you can produce and accomplish and cash in. You taught me differently.

You were the first person who made me feel like I was enough. Many of our friends loudly noted how exhausted they were from making banners and chanting at dozens of rallies, while others flexed their knowledge of political texts like *Philippine Society and Revolution*, their relationship to the masa, and their detailed understanding of ideological rifts among the Philippine left. Meanwhile, you simply invited me along.

You were also the first person in our activist community to invite me to something personal, something that didn't require my presence as a turnout number or my body for a blockade. Through this simple act, you became for me a "super coregulator," someone with whom I felt safe, someone who taught me through your own nervous system how to be comfortable in my own skin. I believe you did this because it was natural to you; it was how you moved through life—looking for enjoyment and creating it for others.

Meanwhile, I constricted my life by avoiding intimate contact, and so often felt down and alone, even when surrounded by hundreds in my activist work. To stay in community, I drew inspiration from a parable that Tara Brach tells about a group of porcupines facing a very cold winter. Some of the group decide to go it alone, and they wander into the forest and freeze. Others in the group decide to huddle close. And though their spines prick one another and wound their closest companions, together they manage to survive.

I stayed in community, but some of the spine pricks I got from huddling were so deep, they scratched at the bone. While I found healing and belonging and purpose, as well as collective impact on critical issues, in some cases I also found homophobia, militance above humanity, and ableist expectations of what you had to put out for the cause. Back then, many of us in the Bay Area radical Filipino American community hadn't yet learned how to distinguish between the martial energy directed at powerful targets and larger systems, and the more subtle negotiations we needed to be in compassionate conflict with one another. Back then, it was not yet the community of practice around healing that it may be becoming now.

You and I commiserated about this. Our relationship became a refuge from the everyday cruelties of the activist world. We weren't "down" enough to shun other people for not being down enough for the cause. In between teach-ins and press conferences and rallies, we could just *be*. We spent hours in silence, listening to music and smoking weed. We goofed around and treated laughter like a drug. We giggled through karaoke contests, competing to see who could be the bigger ham. You always won, with your over-the-top renditions of Earth, Wind & Fire. You were OA, as Filipinos say—overacting. You were maarte, extra, a hilarious drama king who taught me how to laugh at myself and have fun. This for me was a new form of love.

My therapist had explained to me that there are two pillars of self-esteem—feeling capable and feeling lovable. He said these two pillars are key to resilience in the face of trauma. I had always felt capable, and our activist lives fed this pillar. But I had never felt lovable. With you, that started to change.

~

In the spring of 2000, we joined Diskarte Namin, and shortly after, we spent the weekend in Stockton recording our first songs. J's brother had a makeshift studio in his garage, with a utility closet turned isolation booth padded with egg-crate foam. You stepped in there, put on headphones and sunglasses, and shook your head around and sang, like you were Stevie Wonder or some shit (you wish). Your antics softened the steel in my limbs. Your special brand of stupid-silly was a signal to relax the bracing my body still thought necessary to survive. So I was startled when, after putting down your vocals, you invited me to a wedding that very evening in Sacramento. My first thought was *Plan much*? My second thought was *Oh shit, is he actually inviting* me*? Just me?*

At that time, most people didn't invite me to things, probably because I radiated a vibe of not wanting to be included. Also because I was awkward as fuck. I didn't know how to make small talk, giggled nervously at inappropriate times, and failed to laugh at the right times because I often didn't get the joke. I desperately wanted to be included, but sustained social interactions were new to me and I didn't know how to do them. Deep down, I remained convinced that I wasn't worthy of anyone's time, including yours. And so I literally turned my back on people to avoid painful rejection and to pretend that I didn't need anyone—like I was 100 percent okay on my own.

But you saw right through that front. Beneath your shit-talking and goofball antics, you were deeply intuitive. Constricting and avoiding were defense mechanisms that you understood, and you seemed to treat my off-putting vibe as its own invitation.

That night, you drove us from Stockton to Sacramento in your beat-up Toyota Corolla. You were famous in the community for practically living out of that car, crisscrossing the Greater Bay Area multiple times in a day to visit friends, perform odd jobs, or attend a variety of rallies and cultural events, like the one we were headed to that night. I don't remember what we talked about on the drive or if we talked at all. I only remember my bare feet pressed against the dashboard and Sly Stone playing from your car stereo—you were a Vallejo kid through and through.

When we arrived at the Sacramento State auditorium, you dropped some sudden news: you were part of the wedding and would see me after the ceremony. Before I could open my mouth, you were gone. Perplexed, I folded down an auditorium chair and sank in.

Moments later, the curtain drew back to reveal a stage bursting with flowers. The bride and the groom floated to the center. Robed in white, they moved with the dignity of icebergs gliding through water. On either side of the bride and groom were pairs of people draped in dazzling cloth, bearing robust red and orange blossoms around their necks and in their arms.

Lines of people descended the auditorium stairs to join the bride, groom, and entourage on the stage. You were the last in line, but you were not yourself. You bounded down the aisle in what looked like slow motion, in white, flowing pants with a scarlet sash around your waist and large red and white beads around your neck. Down the steps you came, legs bent to bring your body close to the ground, knees rising to your chest, and arms swinging high above your head in commanding sweeps. In this way, you stomped quietly onto the stage. All of you, even the swish of your pants and the sway of your beads, seemed to be moving in water. Your eyes bulged, and your mouth moved open and closed with your cheeks puffed out like a blowfish.

You were embodying Changó—the Orisha deity of dancing, drumming, thunder, and fire, and the groom's personal god. You and several other dancers had created some of the pantheon of the Santería religion and surrounded the Santero couple with their closest deities. You explained this to me as we hurtled back to Stockton doing 90 on the I-5, the full moon bathing your car in watercolor blue.

Now, when I think of a body released from the trappings of trauma, I think of the way you danced Changó. This was truly the essence of you. You were freer than I had ever seen you, animated by spirit and a depth of force that you didn't reveal to most others in your orbit. Your sister, Lyn, later told me that she saw you become this free when you sang onstage in high school. Here you took that freedom to a spiritual plane.

Witnessing this ceremony welcomed a new layer of vulnerability and intimacy into our shit-talking, weed-smoking, karaoke-singing companionship. My neural networks changed from this experience. Though in general, neural plasticity occurs over time, sometimes networks learn quickly, with activation and rewiring taking only a matter of seconds. This was one of many brief but impactful instances in which you helped rewire my circuits toward a new baseline of safety. By inviting me along to this spiritual ceremony, you cracked some of the ice my body had forged to protect myself from harm. You began to show me the transcendence that lies on the other side of security, which means you started to help me get free.

~

That was the summer of 2000. Over the next few years, you would continue to invite me out of my shell through actions as simple as they were consistent: When I sat alone in the corner at band practice, you would come over and show me how to play a new pattern on the kulintang. When I was in a surly mood at a community event, you'd elbow me until I paid attention, then tell me bad Filipino accent jokes about dry bears and copy machines. When I tried to escape from the rest of the world, you would find me and bring me back to this plane.

That's what happened in October 2001 when I came back alive from Fort Funston, the day I almost jumped off that cliff. How did you know that I needed your help? You appeared at my doorstep with your messenger bag slung around your back, and out of its depths you produced a burrito and a joint.

This is also how you showed up in the years to come. During our human rights trip to the Philippines in 2002, you picked up my passport from where I'd dropped it on a jeepney floor and retrieved my wallet and phone from an office in Manila. At an antiwar protest in 2003, you found my keys at an intersection near the Oakland federal building, where they had spilled out of my pocket as I marched. These are the traits that would have made you an excellent paramedic, which is what you wanted to be. You had the instinct to help people. You picked my things up, you picked me up, you wouldn't let me fall. I don't know what I did to deserve it.

At the end of our trip to the Philippines, a teenage friend got the idea to pierce my ear—no ice, no alcohol, just one stud with a metal post sharpened to a point. When she pushed the stud through the cartilage at the top of my left ear, the point hit a nerve and it felt like a nail splintering sentient wood. You let me squeeze your hand so hard, I thought I'd hurt you. But your body didn't break, not then.

~

When you moved in with me in late 2002, the rumor mill ground doubt to dust. Everyone in the community knew it: we were a romantic couple! We snickered and added grist to the mill. We made it a point to attend weddings together where we danced and posed like prom dates. We enjoyed feeding the narrow imaginations that could not envision two cis-presenting heterosexual people in a deep but platonic relationship.

Meanwhile, our reality was this: you were my brother, and you were more. There are no sufficient words to describe our relationship on this plane. All I know is that beacon of light that was your smile and the life preserver that was your outstretched hand both kept me from drowning.

And when it came to romantic love, we helped nurse our respective broken hearts. Whenever a date sucked or a lover failed, we would stand on the stairway landing of our apartment on Florida Street in the Mission, smoking American Spirits and propping each other up while cutting our love interests down.

"You're outta her league; you're major and she's little," I'd say.

"Back at you," you'd reply. "You gotta aim higher. Anyway, he has chicken legs."

"Chicken legs?" I laughed, waving away the smoke you'd blown in my face.

"Chicken legs," you nodded, making squawking sounds and filling two shot glasses with Jim Beam. Long drags on the yosis, smoke rings in the air.

"To us," you said.

Clink went our shot glasses, followed by a slow burn down the throat.

"To us."

~

You and I shared a nervous connection. Our autonomic nervous systems entrained to each other, the neural equivalent of being on the "same wavelength." We knew through our nerves that we were safe together, perhaps because we both kept similar demons and kept them well.

These demons, our traumas, were only vaguely sketched between us, and they brought us together as much as our need to pursue joy. It's not that we "trauma bonded" per se—we didn't need to tell these stories to feel close. Instead, we bonded over a largely unspoken familiarity with ancestral trauma from migration and the intergenerational impacts of war. The cues were small but eloquent: we both knew we had fathers and grandfathers with military history, parents who had cut family ties to migrate, a history of barely suppressed violence in our blood. I believe our neuroception—the sort of sixth sense through which our nervous systems detect threat—sensed these depths. It was the combination of these currents and how we preferred to rise above them through music, laughter, and activism that cemented some of our special connection.

But unresolved traumas can surface as storms. Neuroscientist Lisa Feldman Barrett says, "The best thing for one's nervous system is another person's nervous system." She also says that the worst thing for one's nervous system can be another person's nervous system.

~

Do you remember that night we walked around the Sutro Baths? It was an evening sometime in the summer of 2003. The moon shone like a pearl, its light reflecting on the waters of the ruins. As we walked along the seaside trail, gravel crunched beneath our feet and a chill wind blew in from the ocean. You told me about a friend at work, a psychic, who said they saw a new path unfolding. You thought it a sign that you should finally become a paramedic, something you had wanted to do for a long time. For reasons I didn't completely understand, you had been taking tentative steps toward this goal but had yet to go all in. This was to be your leap of faith. I remember you felt hopeful, as if a burden had been lifted. Toward the end of our walk, clouds obscured the moon, and we could barely see the steps in front of us, the uneven gravel, the sloping ground.

Soon after, we drifted apart. No, if I'm being honest, we didn't drift; I erected a dam to separate my life from the floodwaters of yours.

It started with small things. A tower of canned food I'd shipped from the Philippines became a scant row of two or three containers; the rest had fallen victim to your pantry raids. Lyn later told me that you'd always had an enormous appetite. She called you "Pelican" when you were a child. I guess some things don't change. I'd pour myself cereal and find it gone, except for the one goddamn flake you'd leave at the very bottom of every box. My stash of weed also got smaller until one day—along with my pipe—it disappeared.

I was okay with this at first. You had shared plenty of food and weed with me in the past, especially that night you saved me after the cliff. I knew it was a rough time for you. You didn't have a full-time job, you were on contract as a mover and a warehouse worker for Pottery Barn, you were feeling stuck and hesitant to take the leap to become an EMT.

I just had to be patient. You said you would begin your paramedic requirements soon.

But then you stole money from the band. And at the time, the band was grinding. We did two, sometimes three gigs a week, trying to scrape together enough funds to do a college tour in the fall, a Cuba trip in the winter, and a spring trip to perform at the 2004 Cordillera People's Festival in the Philippines. You'll recall that we bandmates were arguing with one another—a lot—about expenses for studio space and recording. Some of the members wanted to split the gig money to use for living, in some cases to help raise their kids, while others were adamant all funds should go back into the band.

So when I found out the money was gone and that the whole time the rest of us had been fighting about how to use it, you had been spending it on your rent, weed, and lord knows what else, the band was pissed, but I was livid. In the apartment we shared, I stormed into your room and flung open the accordion folder you used to store Diskarte money and receipts. I turned the folder upside down and shook it wildly. White paper fell to the ground along with just a few green bills. I grabbed the bills and counted them—forty-five dollars. That was all that was left of the thousands we had been saving for years.

"What the hell, BJ?!" I yelled. "You've been spending the money this whole time, with me right here? What are we supposed to do about Cordi Day? And Cuba? How are we gonna pay for the studio now?"

I don't remember how you responded. My memory was probably impaired by my rage. I felt personally attacked. I was exhausted from work, from my chronic conditions, from the amount the band had been gigging, which suddenly felt like it was all for nothing. I would have lent you the money if you had asked. Instead, you went behind my back and betrayed my trust, the whole band's trust—our family's trust. For me, our neuroceptive bubble of safety had been destroyed.

"You have to go," I said.

I hadn't yet learned how to "talk it out" or how to be flexible and porous with my boundaries. I only knew the two extremes of relating—letting someone all the way into my heart or shutting them all the way out. I know now that this is the very definition of conditional love.

We suspended you from the band, and you moved out of the apartment we shared. We rarely talked after that. I missed you but never admitted this to you or to anybody. I told myself I had too much pride. Really, I was just too hurt. I felt there was a way to do things. I had my own private code that you had violated.

In the end, we became separated by our deepest trauma responses, the unhealthy survival reflexes still present in our nerves. Your body may have fallen into old patterns that kept you stuck. Meanwhile, I was triggered by feelings of abandonment and being used, which I felt erased our track record of mutual love. The devolution of our relationship became an example of how unresolved trauma can keep us from being our best selves, of how our hurts can keep us separate when we need each other most.

And so, a year later, when you called me asking for help, I was still as angry as if you had moved out yesterday. First, you called my friend Emily to ask her advice as a doctor. You told her you were in pain. She told you to go to the hospital immediately, but you didn't want to. Days later, you called me. This time, you asked for a ride to the emergency room.

"Are you okay? What's wrong?" I asked with guarded concern.

"I don't know. My hands are all tingly."

"How long has this been going on?"

"Weeks. Months, maybe."

I felt a small rush of resentment. Here you were, stuck again. Not taking action and bringing other people down with you. "Well, where are your roommates? Can't they take you?"

"No, no, I need to go right now. I can't even move."

I waved Emily into the room and told her what was happening. "If he can't move, we can't move him," she said. "He needs an ambulance."

"If you really need to go now, then call an ambulance," I repeated to you. Then I hung up.

You died a few days later.

~

The last time I saw you was November 2004, three months before that last call. I missed you—we had barely talked the whole year—and I was moved to throw a karaoke party for you. You and a handful of friends gathered in the apartment I shared with Emily and her husband, Daniel. You crooned and cackled and sang "Reasons" by Earth, Wind & Fire—perhaps one of the most depressing songs ever made—like it was a comedy routine. As your lungs belted, your back arched and doubled over, your legs kicked and squatted, and everyone in the room busted up laughing. We had to throw things at you to make our bellyaches stop. This was the same cheesy charisma that had won you so many fans, from your musical theater shows at Sacramento State to your performances as the front man of Diskarte.

The karaoke gathering was a despedida—literally, a farewell party—the kind you throw in Filipino and Latino cultures when you're not going to see someone for a while. You were going to the Philippines with Lyn and your mother for two weeks. I was supposed to see you when you got back, but life, and then death, got in the way.

~

You died after midnight on February 8, 2005. It was sudden, and no one knew what exactly had happened. Rumors flew: you had overdosed, you were poisoned, you were cursed by the Devil. Much later, after doctors sent a sample of your heart tissue to the CDC, we found out that your official cause of death was myocarditis, a viral inflammation of the heart that can cause sudden death even in young, otherwise healthy adults.

I believe that the trauma you carried could also have been a factor. Like too many other people in brown bodies, including otherwise able-bodied cis men, you may have been marked by unresolved trauma, a pre-existing condition which too often turns sickness into terminal disease.

~

If the sun rose the morning you passed, I never saw it. My eyes were shut the whole day. I slept what I imagine to be the sleep of the dead.

When I awoke, my eyelids were heavy, like curtains soaked by rain. I had been crying in my sleep, and the tears were waterfalls pooling in my ears, cascading onto the pillow and into the space between my collarbones.

I felt in my body the loss of everything you were, not just to me but to the whole community. There would be no more karaoke sing-offs, no more shows with you as the front man of Diskarte, no more workshops where you taught indigenous Filipino instruments, no more driving across the state to help with activist events from San Francisco to Sacramento to LA.

After you died, your communities celebrated you through multiple memorials in several locations. Dozens of your friends filled your apartment. We covered the walls with poster board and pinned the poster board with memories, including photos and testimonials and heartbroken diatribes cussing you out for leaving. The way the community came together to collectively mourn your death was a deeply healing ritual for me. It made all the difference between a tragic loss and a traumatic loss, the difference between what Resmaa Menakem calls "clean" and "dirty" pain.

You and I knew each other for only five years, but you live on in collective memory even after your death. Lyn started a memorial fund in your name to pay for lessons for young artists and to carry on your love of music. And, of course, we threw you a karaoke party, which we called "Legacy." You have left me many legacies, including a vibrant friendship with Lyn, who loved you to the stars. I now understand that she is one of the reasons you were able to give unconditional love to me—you learned it from her. You also live on in the form of questions that inhabit my body-mind: *How am I carrying forward your example of unconditional love? How can I be better at giving this incredible gift that you gave me to those around me who still live?*

I am at least practicing giving it to my child, and I know that if you had survived, you would have been one of his best friends.

Broken Water

In Philippine folklore, one of the most fearsome monsters is the manananggal. A beautiful woman by day, the manananggal transforms into a flying predator by night. She breaks her body at the navel, leaves belly and legs perched in a tree, then flies about in search of prey, wild hair and intestines trailing in the wind. When she finds a suitable victim, she probes their bodies and feeds off their innards with her long tubular tongue. Unborn children are her favorite meal. She likes to suck them fresh from the womb. On occasion, she will use her long, sharp fingernails to excise them from their mothers' bellies before devouring them whole.

~

"When are you gonna have kids?" my coworker Violet asks me. We are sitting opposite each other in the waning afternoon sun that stretches across the Berkeley nonprofit office where we both work.

It is the spring of 2001 and I'm a year into this job at a Filipino arts and ecology center. I'm also a few months away from quitting because of intense pain in my neck and arms. My head is buried in a stack of articles about Philippine biodiversity, and I've been massaging my aching wrists as I cull information for the next newsletter. Violet, who manages the organization's art gallery across the street, has snuck over to take a break while the bosses are away.

"Do you feel ready? Like, sometime soon?" she presses.

I look back at her blankly. It takes me a minute to orient to the certainty of her question; the when, not if; the assumption that having kids is as inevitable as having lunch and something you plan into your day.

"Oh," I finally say. "I don't think I'm having kids."

Now it's her turn to look at me blankly. "What? Why not?"

I'm annoyed at her probing. But I consult my belly, the area where my womb is supposed to be. *Hello, are you ready?* I ask it silently. No response. So I shrug. "I just can't see myself having a baby."

Violet throws her curly black hair over one shoulder, shifts in her chair, and leans toward me. "Oh, I think you'll change your mind when you meet the right person."

Again that certainty, which irks me into as much counter-certainty as I can muster: "Mmmm, I don't think so."

Violet flattens her lips in a disapproving look, then turns the conversation to other topics.

At the time, this was my truth. Why would I want kids? There was no biological urge in my body, and at that stage, I didn't yet feel disgust. I touched my belly again. Where my womb was supposed to be I felt a numbing hum, like a washing machine stuck between cycles or like a dead space where my body might detach.

~

The manananggal has found its way into Filipino pop culture through movies, plays, books, and even a Netflix series based on the graphic novels *Trese*. In most of these portrayals, the manananggal is the antithesis of the glorified mother figure. She does, after all, leave her reproductive organs behind while stalking her prey at night. And she does, after all, prefer slaying unborn children to giving them life. In many ways, the manananggal is the foil to the ultimate mother in colonized Filipino culture: Mother Mary, she of the Immaculate Conception.

My middle name is Mary. But I have identified less with my namesake than with the manananggal.

~

As a heterosexual woman or, more accurately, a nonbinary femme androphile in a cishet-presenting body and relationship, I was subjected to many variations of this "When are you having kids?" conversation, especially from my partner's family and my own. Filipino Catholics and Puerto Rican Catholics alike assumed it was only a matter of time before nature took its course.

A queer colleague dubbed it the "When are you going to breed?" conversation, which is exactly what it felt like. I resented the heteronormative assumptions and the Catholic directives behind the question. I disliked the patriarchal ableism of being treated like a vessel for reproduction. And I hated that my coworker Violet, for the most part, turned out to be right.

I met my partner, Juan, and eventually changed my mind about wanting to have kids. When I first laid eyes on him, I was sure I had seen him before. We met at our mutual friends Emily and Daniel's wedding, where we danced salsa for hours. In the months that followed, we had jam sessions for dates, he sent me letters with drawings he had done at live music shows, we went to art openings, he watched me sing at jazz clubs, and I wrote him bad poems. Our lives became a tornado of artistic creation. Juan moved in with me, and a few years later, in 2007, he joined Diskarte Namin, the band I had been in with my friend BJ. All of this creative activity led me to believe that we could also one day create a child. For the first time, the thought of carrying a biological baby became an intriguing possibility.

But chronic pain had other plans. One night at our apartment in San Francisco, I trembled in bed next to Juan, trying to swaddle myself in sheets, trying, I suppose, to create a body-size tourniquet to ease the pain. Lying on my side, I felt the arachnoid cyst in my lower back insist itself into my spinal fluid. I imagined hairs on its spidery legs scratching at my nerves like nail files.

I had to disrupt this untenable state; I had to do something that would bring me relief. I looked at the plate-glass window of our bedroom, how it reflected the sickly streetlight, and thought, *Wouldn't it be nice to go through that glass? To throw my body against it and be sliced and to fall, my nerves severed, pain signals trailing in the wind?*

Imagining this made me feel a bit better, and instead of acting on my vision, I tightened the swaddled sheets around me like a cocoon. I told Juan—in the way that one might confess they are ruined—that I didn't think I could have kids. "I just feel like the pain would destroy me," I said. The words came out like a whimper, and I felt disgusted at myself for being so weak.

I will never forget how simply he replied: "I'm not with you to have kids. I'm with you to be together with you."

I blinked back at him and began to sob. "I'm sorry I'm such a mess," I said through tears. But I felt a surge of love and gratitude. Freed from obligation, the possibility of having a child with Juan grew into a longing. I seeded this longing at the back of my heart like a spring bulb tucked away for the winter.

In the meantime, pain was still my primary relationship. It occupied much of my thoughts, sucked away most of my energy, and ate up a lot of my cash. I was fulfilling the logic of pain, whose purpose is to draw attention to itself so that corrective measures can be taken.

Over the course of the next four years, I threw the book of corrective measures at my rebellious body. Chiropractic, acupuncture, spinal traction, homeopathy, yoga, somatic therapy, sacro-occipital therapy, craniosacral therapy, physical therapy, electrical nerve stimulation therapy, EMDR, and heavy doses of medical-grade weed and muscle relaxers. I spent the majority of my income on these therapies—which were either partially covered by insurance or not covered at all—and what was left went to rent while Juan paid for food and everything else. Meanwhile, I devoted mental and emotional resources to cobirthing a nonprofit organization dedicated to media justice as well as starting my own consulting business.

This was my litany of creation. Taking up leadership in movement work activated a healing sense of agency and aligned with my desire to appear strong, effective, and invulnerable. Integrating a battery of somatic and psychological therapies began to take the sharpest edges off my pain. My arms no longer felt burnt to a crisp, breath could enter my lungs without battle, and the rebar-like pain in my hips was filed down to a dull ache. When I touched my belly, it still felt numb. But pain was no longer my singular identity. It became something I walked with in some parts of my body rather than the puppet master that controlled all of me.

One morning, I went to the kitchen to make tea, and when I saw the message on the label, I smiled. It read: measure yourself not in how much you've achieved, but in how much you've overcome. I was proud of what I'd overcome. I had once felt like the walking dead, so fully had I embodied a traumatized state of disconnection and despair. In this context, insomuch as I could feel, perhaps the residual pain was actually a sign of living.

Nine years before, I had reached a breaking point and tried to take my own life. Now, through a combination of intensive professional therapies and relational healing practices, I embodied a new state of relief and connection, which created space for even more growth, and potentially more life.

In this context, we received auspicious news: Juan had been accepted to a fellowship in community development in Puerto Rico. This was a dream for him. For years, he had been wanting to move back to his birth island, and now he had the perfect opportunity to contribute his skills back home. We were both elated.

As we packed up our things in the Bay and made plans for our new lives in San Juan, we discussed how moving to the island could start a slower chapter in our lives, during which we could try to conceive a child. My pain was better, the timing seemed right, the way looked clear.

~

A week before we left for Puerto Rico, I had my last therapy session with Bret Lyon, whom I had been seeing weekly for seven years. When I told him that Juan and I decided to try having a child, the announcement gave him pause. He had seen me cry in most of our sessions—torrential tears that formed the wake of what he explained to be a deep attachment wound. He himself had been subject to emotional neglect. He had once briefly shared that he chose to be child-free because he did not want to pass on the traumas of his own upbringing, and I related to this 100 percent. But seven years of therapy had brought me to a different place. After a moment of silence in which he appeared deep in thought, Bret finally said, "It's a surprise, I'll be honest. But I can see how this move will help. I can see how raising a child could be healing for you."

~

In some depictions of the manananggal, she disguises herself not just as a beautiful woman, but also as a healer. She travels from village to village, tricking people with her false cures, then at night attacks the very same patients she has treated by day. In other words, the manananggal is not to be trusted, and healing can be a double-edged sword.

~

Juan and I settled in the colonial village of Old San Juan. Our apartment had a small balcony with a partial view of San Juan Bay. I relished opening the balcony doors each morning to watch village life pass by. I saw our landlord place bowls of food on the sidewalk for the bevy of stray cats that wandered the cobblestoned streets. Families strolled the sidewalk while governor's guards stopped to talk to the young man selling piraguas on the corner. On some days, I could hear the flower vendor sing, *Aaaaaaazucena!* as he walked the streets, holding an enormous bunch of long-stemmed white flowers under his arm.

In this latticework of life-giving culture, Juan and I committed to trying to conceive a child. It wasn't all a romp in the park. I religiously charted my ovulation cycle, and we tried the "natural way"—at least, the natural het way—for two years. When this didn't work, I went through a variety of tests, some of which caused my chronic pain to flare, only to find that one of my fallopian tubes was blocked with scar tissue and the other coiled like a Slinky. I imagined the trailing intestines of the manananggal, hastily gathered and smashed back into the gut to reconnect her body in time for sunrise. My fallopian tubes were similarly jammed around my viscera. There would be no eggs making it through those passages.

The fertility doctor offered the choice of a laparoscopy to clear my tubes so we could try again "naturally." The alternative was a direct path to in vitro fertilization (IVF.) We decided to go straight to IVF. In Puerto Rico, the process costs a third of what it can cost in the States. And I was working a full-time consulting load, getting paid generously in the process. IVF was not only financially viable for us, it would also allow me to skip surgery that could possibly lead to more scar tissue and pain.

Once we'd made the decision, something in my soma shifted. I began to be plagued by visceral doubt that went beyond my fears of increased pain. I was seized with concern about conceiving: How could there be room for a child within a body such as mine, within a life so organized around pain? What if my pelvis separated like the two halves of the manananggal? What if I survived pregnancy and childbirth only to lash out and beat the newborn until the problem of its draining existence was resolved?

~

Within months, my doubts about becoming pregnant had morphed into delusion. Each thought of a baby inside me sparked an urge: I wanted to grab a kitchen knife to cut out my womb. I saw pregnant women in public and was disgusted by the sight of their bellies—I wanted to protect them from the vulnerability of being so obviously with child. When I looked at them, a terrible vision overcame me: I saw a body-size mandolin slicing the bulge off their midsections. I closed my eyes against the horror of this imagined scene, but I also felt a strange sense of relief—the mandolin, I thought, would rid them of the debilitating growth, so they could run and hide for their own good.

Something in the wraithlike possession of me, the walking trauma in me, was making me orient toward having a child in a way that was frightening. This terror began to show up in my dreams. One night, I lay awake in bed till I saw light bleed through the windows. When I finally fell asleep, I dreamt that Juan lay on a table with his legs restrained. Ants swarmed his body, and he swatted at them and scratched his limbs until he bled. The scratches grew into gaping sores. Alien-like tendrils sprouted from the wounds. The tendrils consumed him, and as they approached his face, I startled myself awake. I lay in a pool of sweat. I had scratches on my forearms. I must have raked them with my fingernails in my sleep.

I became convinced that I was the manananggal. I was not meant to have a child because I would prey on this child and bring suffering onto others as well as myself. Perhaps trying to conceive was simply bringing me to my final path and purpose: to kill a monster. What could these violent visions be but the preoccupations of a beast, the potential energy of an aswang? Slaying the monster, I thought, was the only way to end transgenerational cycles of pain. I had given up once, but this time I believed I would succeed. It could all end with me.

~

To kill a manananggal, you must be ready to confront her face-to-face, or you must be willing to journey in search of her lower half. If you are brave enough to confront the manananggal, you can slay her like you might end a vampire: by driving a sharpened bamboo stake through her back. If you are persistent enough to find her lower half—perched on a tree limb, stuffed in her home closet, or hidden in long grass—you can prevent the manananggal from reattaching by smearing salt, vinegar, crushed garlic, and ash on her torso. If she is unable to make herself whole after the hunt, she will die before dawn.

~

I had been at the edge of a cliff before, and I recognized suicidality when I felt it. I knew I needed help. So I opened up about my feelings to two dear friends, both of whom had been part of my community in the Bay. Lakshmi and her four-year-old daughter, M, came to Puerto Rico all the way from Bangalore. Lakshmi was my first friend in San Francisco. Her no-bullshit attitude and the way she spoke with a warm, throaty voice and the animation of her entire body warmed me to her immediately. We bonded over being women of color in a white liberal magazine industry, and also over late-night drink and smoke fests at various downtown bars. In San Juan, I recreated a bar in our apartment; we sat on the balcony drinking wine while her daughter played beside us, just out of reach of secondhand smoke.

"I feel pretty messed up," I confessed to her between gulps of wine.

"Don't we all," Lakshmi replied in her throaty voice as she tilted her head upward and exhaled smoke from her Camel.

I shared that I'd been having suicidal thoughts for the first time in years and that the prospect of having a baby was somehow driving me to a place where, rather than try to bring life, I felt I ought to take it.

M stepped over our feet to grasp the iron bars of the balcony, and I noted how a fall over the top rail could be deadly for her. Lakshmi must have noticed the same thing as she stood up immediately to usher her inside. As I watched M play on our futon, I found it hard to imagine that I could create a whole new being such as this. M petted our dog and giggled in Kabu's face, and I felt a pang of hope.

"I look at your daughter and I just want to be able to do this without any drama," I told Lakshmi. "But instead, I have all these terrible thoughts. I keep thinking about walking into the bay and never coming out."

We looked at the dark slice of San Juan Bay, visible from our balcony just beyond the streetlamps of the governor's mansion. Some moments I thought it would be better for all if I walked toward that inkblot to discover no page beneath, but instead an ever-yielding, descending soft sand. No one would get hurt but me. I could be a Filipinx Edna Pontellier, and at least the water would be warm.

Lakshmi put her hand over my knee. She smoked, and we drank, and we talked late into the night because she knew that talking is medicine. Being in the presence of one who loved and listened was healing in and of itself. No judgments, no overreactions, no trite solutions. Just flow.

For months after Lakshmi's visit, whenever I sat on the balcony and looked out at the bay, I would still think about walking into the water, but each time, I could see myself staying closer and closer to shore, until I could envision dipping a foot in and then turning away.

My friend Amy flew over from Oakland later that summer. A tender soul and a radical thinker, Amy had been a mentor to me since our years doing racial justice work in California. She is also one of the best listeners nature has ever made. One evening, we snuck into the hotel across the street to use the hot tub. We soaked in the bubbling water, dipping in and dipping out to cool off on the terra-cotta ledge.

"It's not like I'm afraid of gaining weight or having swollen ankles," I told her.

"It's the pain," she said.

"It used to be, but I'm not even afraid of that anymore." I swirled the steaming water with my foot. "Now it's about the thought of something growing inside of me."

I scooted down from the ledge and submerged my body up to my shoulders. I was ashamed to admit this, but I also knew I could tell Amy anything. "All these mothers I know, they talk about the glow and how amazing they feel while pregnant," I said. "And honestly, to me, it just sounds gross. It's like—a violation."

Amy nodded and encouraged me to continue talking.

"When I think of being pregnant, I basically want to throw up."

"And not like morning-sickness throw up," Amy said.

"Exactly. More like awful, disgusting, poisonous throw up. It just seems like such a bad idea, like, why would I make myself that vulnerable?" My nose wrinkled at the thought that I would be even more fatigued than normal when my belly got big. It was abhorrent to me, for some reason, that others might offer their bus seats to me. That they might look at me and smile with a combination of tenderness and pity.

My lip curled at how much energy would go toward growing this new life at the expense of my ability to fight or flee.

Amy pulled herself out of the water to sit on the ledge. When I shared these thoughts with her, they sounded crazy to me, but I knew that Amy respected this type of crazy. Steam rose to veil her legs as she listened attentively, affirming with nods and yeses, encouraging me to expel what I'd thought too toxic to share.

~

Encouraged by my discussions with Lakshmi and Amy, I took it upon myself to find out more. I googled "disgust." I learned that disgust is a primal reaction and that people with PTSD sometimes exhibit stronger feelings of disgust than people without. *Well*, I thought, *as someone with complex PTSD, I guess it makes sense that I feel complicated disgust*. But why would the thought of pregnancy suddenly trigger such strong feelings?

I read more and learned that there are three possible facial movements we make when we find something disgusting. We open our mouths and stick our tongues out, which is a survival reflex to expel harmful substances from our bodies; we wrinkle our noses, a defensive reaction to noxious smells; and we curl our upper lips to signal interpersonal disgust, which is related to contact with fearsome or otherwise undesirable individuals.

This last fact made my eyes widen in recognition. When I thought about becoming pregnant, my upper lip curled. My body treated the very idea of a baby inside me as a fearsome and undesirable individual. Even though part of me longed to have a child, and Juan and I were moving forward with IVF, another part of me reacted to the vision of a fetus as something noxious, something worthy of being excised by long fingernails and expelled.

I googled "disgust of pregnancy" and found a study about tokophobia, defined as a "pathological fear" of pregnancy and/or birth. One in ten women may experience tokophobia to the point that they delay pregnancy, if they can, or avoid it altogether. I was amazed that this terrifying feeling actually had a name. Why had I never heard of this before? It's a phenomenon so unknown that none of my therapists had ever mentioned it. And even now, more than a decade after this study was published in 2012, my word processer does not recognize the term. Instead, it keeps autocorrecting "tokophobia" to "homophobia."

I now believe the tokophobia symptoms that I experienced are some of the clearest evidence that I carry transgenerational trauma in my body. With every wave of disgust, I felt the potential energy of war in my nerves. I sensed the muscle memory of running, hiding, striking. I felt my grandmother, how she may not have wanted a fourth child, especially one that may have been born from rape and war. I felt my mother and what may have been her reluctance to have a third child, to subject her body to this burden again. I felt my prisoner-of-war grandfather, his lost body found in mine, and how in his soma, there was no biological possibility of pregnancy. I remembered my grandmother's stories of his torture, of how his belly bulged from water forced into his stomach, then expelled by the boots of Japanese soldiers. I became weary from the feeling of fighting endless wars, despair as to what end, and conviction that a new life would only perpetuate this cycle of battle and grief.

In other words, my body treated pregnancy like a traumatic outcome of imperial war. Perhaps it wasn't trying to expel a fetus, but instead was trying to expel the toxic imprint of colonization in my nerves.

~

The manananggal existed before colonization in the Philippines, but she became a more prominent mythical figure during Spanish rule. Early on in the era of the Crown, Spanish friars referred to the "magtatangal" as a "he." But much like male doctors in Europe cast midwives as female witches to undermine their influence, the Spaniards may have regendered the manananggal as female to better associate the monster with the babaylan in order to diminish the babaylan's power. What better way to replace reverence for indigenous healers with reverence for the Catholic Church than to turn a dignified female (or nonbinary) figure of well-being into a dreaded demon of death?

~

From my research, I pieced together a theory: repeated exposure to the violations and violence of colonization may have led my ancestors to interpret everything around them as threats. This may be how they survived long enough to have children. I believe I inherited this hypervigilance to the point that this adaptive response became maladaptive in my environment of relative peace. With regards to pregnancy, epigenetic markers may have disrupted my normal process of threat discrimination such that I regarded even a developing fetus as dangerous.

This makes some evolutionary sense. The polyvagal theory and social neuroscience explain that our most basic neurological circuits are set up to regard each other as potential threats. This is a reflex that does not go away; rather, it becomes inhibited in healthy individuals as we grow.

Our instinct to react to others as potential threats is much like the grasping reflex. Babies reflexively grab when something is placed in their hands. Adults don't because this reflex is inhibited to allow for greater range of motion as we grow. Researchers know this grasping reflex is overridden rather than erased because it can return in elderly people, when cortical functions become impaired due to age. Similarly, the threat response reflex doesn't go away, but in healthy individuals who find themselves in stable environments, it can become inhibited by what's called the vagal brake. The vagal brake is a theoretical circuit responsible for allowing social behavior to take priority over self-defense in conditions of relative safety.

When it came to thinking about having a baby, I was, essentially, uninhibited. I was operating with a vagal brake worn down over centuries. My parents and my grandparents probably needed this primal instinct to react to invaders as a threat in order to survive. This is not just a blood inheritance; it is a collective inheritance: Filipinos as a people forged from colonization have likely scanned for threat and been poised for defense for almost four hundred years.

My nervous system was trying to do the right thing: protect me from a fetal-size germ that could debilitate my overall health, break me from the inside, and render me less fit to survive.

Imagine my relief when I learned that my disgust and delusions in fact made some scientific sense. I went from thinking of myself as a monster—the dreaded manananggal—to embracing myself as a fierce femme just trying to survive. I began to accept myself as a mammal with primal neural circuits lacking inhibition, and a human being in a continual process of adaptation. I began to view a potential new life inside me with more excitement and compassion.

~

Over the course of the next several months, including four months of an IVF cycle, I confronted these traumas through tender heart-to-hearts with Juan, through more conversations with friends, through research, and through bodywork in water, where motherly figures conducted craniosacral sessions in a saltwater pool. I also healed through discussions with my incredibly supportive team of IVF nurses and doctors.

When Juan and I got the positive pregnancy test back from the lab, we screamed and hugged in a parking lot in Santurce. The first cycle had actually worked, even though the statistics were not highly in our favor. A few weeks later, we got to see the fetal heartbeat at our seven-week ultrasound. Instead of disgust, I felt relief and a warm sense of pride.

Every day for nine weeks, Juan, sometimes with the help of others like our good friend Spenta, plunged a 1.5-inch needle into my bottom. With these daily progesterone shots and the confirmed pregnancy, I braced myself for intensified pain and the return of terror and disgust. Surprisingly, what followed was tremendous physical relief. Relaxin and other pregnancy hormones seemed to interrupt pain signals to my brain, and structurally, my growing belly pulled my lower back into a natural curve I never had. The rest of my spine also opened up in a new way, untwisting like the structure of a tree extending toward the sky. Though I experienced issues like recurring urinary tract infections, near narcolepsy, skin rashes, and nasty hay fever, for the majority of my pregnancy, my physical body pain was miraculously gone.

~

A manananggal is born from a woman who has not been fully destroyed by another manananggal. Or, a manananggal is born from a woman who ingests a baby chick from the throat of an elder manananggal. The chick enters her body and is said to sit in the pit of her stomach. There, the chick feasts, eating away at her entrails.

~

Though my pain was on pause, the waves of disgust did not disappear. Instead, they morphed into new emotions. I became afraid of matriphagy, of the growing fetus eating up my body from the inside out. I also became afraid of the opposite, of my own cannibalization of the baby-to-be. I thought labor might throw me into a state of annihilation. I had a vision of crumbs on a hospital bed. The crumbs were what was left of the baby. The smooth muscles of my vaginal wall had clamped down and pulverized its skull.

~

The shadow of the manananggal continued to hang over me. This was my state of mind and body heading into labor and birth. Battle mind. Survival body. I felt ready to detach from my lower half like an ejector seat, if needed, to abort the mission in order to survive. I believe I was also experiencing a degree of gender dysphoria, which has since influenced my identity as a nonbinary femme. Although I no longer felt disgusted or vulnerable, I still didn't feel psychologically comfortable as a pregnant person, nor did I fully identify as a "mother."

In many ways, being uncomfortable with pregnancy and birthing makes sense. The United States has the highest maternal and infant mortality rates of any developed country. Black Americans, Native Americans, and Pacific Islanders experience two to three times the rate of maternal and infant death compared to white women in the US. People of color, poor and working-class people, disabled people, immigrants, and queer and trans people are disproportionately affected by many other issues, including lack of prenatal care, birth trauma, higher rates of Caesarean sections, low birth weight, and preterm birth.

I knew the statistics. But I also felt ready to take on whatever birth challenges might come. We prepared for a home birth in Old San Juan. I wanted to deliver in a tub of warm water and have my baby swim into the world. But well into my second trimester, it became obvious we could no longer stay. Juan's fellowship came to an end, and there were no prospects for sustainable employment in Puerto Rico's colonial economy. As much as we wanted to raise our child in the life-giving matrix of Juan's home island, we were forced to leave.

We moved across the hemisphere to Seattle to be near my brothers and their families. Shortly after we arrived, I signed up for an eight-week-long childbirth class, figuring Juan and I needed all the help we could get. I had no idea at the time that the teacher would be Penny Simkin, one of the founders of the modern American doula movement, and a nationally recognized leader in trauma-informed birthing.

Penny introduced us to her velvet vagina puppet. "See the baby's head, it comes out through here!" she said gleefully. A plush baby head popped out from between the velvet labia and moved around as Penny waved her hand. We laughed. If only birth were this simple.

She taught us an exercise to help the baby crown during transition. "I want you all to repeat after me," Penny said. "Pa, pa, pa."

"Pa, pa, pa," we all complied.

"You've seen what birth looks like on TV, right?" Penny asked. "All that teeth grinding and primal screaming and the doctor yelling at the mom to push?"

We nodded.

"That's hogwash," Penny said. "When the baby's head crowns and you're in transition, it's time to *stop* pushing. Your contractions have gotten you that far, and it's time to ride the wave. Try it again: pa, pa, pa."

"Pa, pa, pa!"

"Saying this will help you remember to stop pushing, and the breathing that comes with the syllables will allow your inner muscles to do the work. Trust that you have already done most of the labor."

I, for one, had been ready to push and scream until my guts came out. But this gem of knowledge helped me prevent severe vaginal tearing.

Penny's class disarmed me. What's more, Penny volunteered her time to help me create a trauma-informed birth plan. Juan and I met her at her home in Capitol Hill, and there, sitting on her living room couch, I disclosed all my fears about becoming catatonic, plunging into a hole of despair, and trying to kill the baby as it made its way through my vaginal canal. Juan held my hand through it all. Penny nodded encouragingly the whole time—she had probably heard some version of this before. She helped me translate these fears into a birth plan with simple requests for the midwives and nurses who would make up my birthing team. The plan explained what I might not be able to share in words during labor:

- I may have strong emotional reactions, so I will need help breathing, moving, staying warm, and feeling nurtured and supported.
- I'm more worried about emotional triggering than I am about physical pain. I want my birth team to not mistake any emotional reactions I might have (like crying or even sobbing, shaking, etc.) as a sign that I need pain medications, and to instead let me move through the emotions with support.

When it came time to give birth, I felt prepared but scared. Juan and I went in for a routine thirty-eight-week checkup. During the appointment, the ultrasound tech left me lying on the table with my belly exposed and still covered in goop. When she returned what felt like eons later, she had a nurse with her.

"We have to induce you today," the nurse informed us.

My amniotic fluid was low. My placenta was apparently not working properly and so the baby's growth had slowed significantly since our last checkup at thirty-two weeks. I suddenly felt that all the preparation of childbirth class had been for nothing. I imagined the manananggal's long fingernails cutting the baby out of my malfunctioning womb.

We checked into the Group Health birthing center, and I was immediately attended to by a kind nurse named Anya. The rotating team of midwives took turns initiating and monitoring a gentle induction process, starting with an overnight balloon catheter and ending in slowly increased doses of Pitocin, a chemical that initiates contractions.

The next day, my dear friend Lisa flew into town to act as my doula. Like Lakshmi, she was one of my first friends to have children and is a fierce Taurus—her earth energy has always been grounding to my watery fire. Early in the induction process, I started to bleed. Lisa had gone out for a run because it seemed the contractions were progressing slowly and everything was under control. I was eating leftovers from a full meal (they allowed you to eat real food at this birthing center) when all of a sudden, I thought I had peed on the floor. I looked down and saw a puddle of red. Were my visions of crushing the baby coming true?

Anya rushed to bring in Sarah, the midwife on duty, who asked me several questions while staring at the puddle. "Hold tight," Sarah said when she was satisfied that no more blood was coming. "Sit down and drink some water. I'm going to go consult the ob-gyn."

A few minutes later, Sarah returned.

"You're experiencing a placental separation," she said, "which means the placenta is beginning to separate from the uterine wall. It's all right, though," she rushed to add when she saw my face. "It must be minor given the amount of blood that's come out."

I had stopped breathing but started again. A placental separation seemed far less scary than my uterus committing prenatal infanticide. Unless this was perhaps its way of accomplishing this goal? I kept these thoughts to myself and went back to picking broccoli florets and fries off my plate while staring at Sarah to better hear what would come next.

"I thought we might have to go straight to a C-section, but Dr. Wang said we could try this first. I'm going to do an amniotomy, which means I'm going to manually break your water. It won't hurt."

I stopped eating, realizing that what she was describing was a last-ditch effort and that I needed to be prepared to go into surgery if it didn't work. All I said was "Okay."

"Dr. Wang said that this should move your baby's head downward, and its head could plug up the blood, and you might be able to proceed with a vaginal birth."

"And that's not dangerous for the baby?" I asked.

"No. The baby will be fine. And we'll continue to monitor its heart-beat."

Sarah put on a glove with a tiny hook and had me lie down on the hospital bed. I imagined a small waterfall pouring out from between my legs, and my baby's head moving downward like an elevator settling onto the ground floor. I'm fairly sure that in most other hospitals, I would have been sent to an emergency C-section the minute someone saw blood on the floor. C-sections are valid, often necessary procedures, and sometimes even preferred forms of birth. But because of my own birth trauma and chronic pain, I was hoping to avoid one at all costs.

After several hours of contractions with no more blood, the next midwife on duty checked my cervix dilation and predicted I had about eight hours to go. Throughout labor, Juan was tireless in his use of an iced can of coconut water, which he rolled over my back and sacrum when each contraction brought waves of pain. After another hour of contractions, the area of my lower back that houses an arachnoid cyst felt as if it was about to tear.

"I can't do this," I cried.

Lisa grabbed my hand and widened her eyes till they magnetized my pupils to hers.

"Look at me," she said. "You've got this. You can do this, you *are* doing this."

I don't remember what else she said; I only remember the iron grip of her fist and how it anchored me to her and to my birth team and to the present. She refused to let me be swept away by the monstrous myth that I was alone.

In another hour, the baby took matters into his own hands. At 10 p.m. on May 1, he pushed his way down my vaginal canal like a cannonball. I had been worried about crushing the baby, but the baby made clear by this last move that he would be the protagonist of this story, not the victim.

I had made a birth playlist, and the songs had been cycling throughout labor. The baby happened to make his appearance to Journey. "Don't stop believing, hold on to that feeling," wailed Steve Perry while I said, "Pa, pa, pa" to the beat. As I *pa, pa, pa'*d, a ring of fire tore through my labia as the baby's head crowned.

His body slipped out of my vaginal canal, and it felt like having the biggest and most satisfying bowel movement you could dream of. The pain of crowning gave way to a rush of triumph. I cried out, "My baby!" just as I heard him wail for the first time. I had delivered on all fours, and so I reached through the tunnel below my torso for the small body that had just exited my own and felt a soft wet matte of hair on a squishy head.

An upswell of water came from my chest, welling to my throat, and finally pouring through my tear ducts. This deluge of resolved grief, another birthing.

My child's body had created a seal with my own. Skin on skin, skull on cervix, blood mixing with amniotic water. This physical contact stopped the bleeding, and he proceeded down my vaginal canal to be born the same day my grandmother was born 104 years before. We named him after her—T. Simeon, the great-grandson of Lola Simeona and Lolo Juan.

We say that water breaks during birth, but it's not that it breaks, it becomes released. When water released from my ruptured membrane, it brought with it the cascade of healing my therapist had predicted. The moment I held my newborn in my arms, I felt a charge of dimensional connection: to him, to the length and width and depth of my body, and to a realm beyond my being, where instead of visions of annihilation, I saw a flash of gold. In my mind's eye, my mother and my grandmother and other ancestors lined up in a row with their hands cupped. My body, still bleeding and newly stitched from a vaginal tear, nonetheless felt a wave of divine gratitude for all those who had delivered me to this moment through their own journeys of letting water flow.

Shortly after birth, we had a party in our hospital room. Juan's parents, my parents, my brothers, and their families all came to celebrate the arrival of T. Simeon. I stood up to embrace everyone, saving my greatest energy for bear hugs for my birthing team. Together we birthed not one but two whole new beings. My baby and I were not yet attached by love, but we were at least not at odds with each other. Instead of predator versus prey, monster versus child, we were transformed into a conspiracy and bonded through the release of water. And that was a good start.

V
Neuromimicry

- The process of drawing inspiration from nervous system design and applying it to the way we care for one another and our environments.
- Where our journey ends in reflecting and envisioning a trauma-wise future.

A ghost
hums through my bones
like Pan's midnight flute
shaping internal laws
beside a troubled river.
I love this body
made to weather the storm
in the brain.

—YUSEF KOMUNYAKAA

Rupture

I have a newborn in my arms. This warm bundle of baby is like a sack of air freshener; he smells of shampoo, diaper cream, and that unique odor that infants have—an addictive combination of sweet jasmine and fresh-baked biscuits. I feel tender toward this creature, which is a relief.

Little T, as I'll call him here, is swaddled like a Mission-size burrito. He has no eyebrows or eyelashes, just small speed bumps where the skin lies smooth over brow bones and a line where his lid meets lower eye, uninterrupted by any trace of fringe. His ear canal holes are pinpricks. He looks like a naked mole rat in human form, and I'm overwhelmed by his vulnerability.

What an incredible place to be.

In this moment, Little T and I are in our fourth trimester, the first three months after birth when newborns behave like they're still in their birth parents' bodies. He is in transition from the dark safety of the womb to this new world with its glaring lights and unpredictable sounds. My role, I have learned through copious reading on infant development, is to recreate the safety of the womb. I am to shush and rock him, soothe him and rest him on my chest where he can hear my heartbeat and feel the rise and fall of breath in my lungs. These interactions help form secure attachments between infants and caregivers, and can help teach babies' nervous systems how to regulate themselves.

What an incredible place to be.

I break our mutual silence with a lullaby, a slow version of Stevie Wonder's "You Are the Sunshine of My Life." Juan and I sang this song to Little T when he was in the womb. Now he seems to recognize it even among the din of this new place. He coos and gurgles in response, and I giggle with disbelief and delight.

Are we forming an emotional bond? I can sing lullabies till the cows come home, but that can't be enough. Will I really be able to form a secure attachment with Little T when I never had one with my own parents? Can I help him learn to regulate his nervous system when my own is so frayed and on edge?

I think of what Pueblo psychologist Eduardo Duran says: that all of us carry a kind of soul wound, a sluice of grief and anger that springs from generational loss. Some of us carry this wound lightly, some so profoundly, it threatens to rupture us from the inside out. Duran writes: "In this moment if you begin to heal yourself, you are healing all the hurts of the ancestors you carry, in this moment you're healing for the next seven generations. What an incredible place to be."

I am full of gratitude for reaching this place, for the opportunity to bond with Little T. Yet even with as much healing as I have experienced, my soul wound remains tender and grieving. Or else it churns like water weaponized by a typhoon. Both emotions are so full of generational momentum that I know I must work to rupture the cycle that would otherwise continue through him.

~

Living with trauma-related anxiety, C-PTSD, and chronic pain is an exercise in knowing my limits. Every day I manage a low body budget easily drained by the triggers of everyday life. While crossing the street or driving to the post office or picking up groceries at the store, I scan for potential accidents at such a level of hypervigilance that stress hormones flood my body while performing these routine acts. In crowds of people, I am constantly looking for signs of trouble, scanning faces, watching hands in pockets, and if I do experience a racist or sexist microaggression, I respond as if a gauntlet has been thrown down for a fight to the death.

This is my version of the cycle behind the illness metaphor of spoons. Those of us living with chronic illness and disability, visible or invisible, often operate with finite energy, or a limited number of "spoons" in our drawer. Each activity we participate in uses up spoons. In the case of people with C-PTSD, one or two normal activities can drain our spoons because it takes so much energy to be chronically prepared for disaster. Once we've used our last spoon, we have to rest to replenish the supply or borrow against future spoons, or else there are immediate consequences. For me, these include pain flare-ups that land me in bed for days, fatigue that necessitates sixteen hours of sleep for three consecutive nights, and angry behavior toward the ones I love.

Parenting with trauma-related anxiety, C-PTSD, and chronic pain is an exercise in stretching my limits—while still operating with the same amount of spoons. The thing with parenting with C-PTSD is that normal baby behaviors like whining, crying, feeding, pooping, waking, and the seemingly endless cycle thereof can grate on anyone's nerves. But for me, C-PTSD, as well as the related exhaustion of chronic pain, can amplify this normal grating to a state of slash-and-burn.

One afternoon when Little T was about two weeks old, Juan and I were attempting the three-ring circus of tube-and-syringe-aided breastfeeding. My milk flow was low, and anyone with low milk production faces a frustrating catch-22: the more the baby feeds, the more milk will come, but the baby won't feed if you're not producing enough milk.

To teach Little T how to nurse and also to increase my milk supply, we had to give him a quick hit of formula while breastfeeding. Sitting on a glider chair, I would place a plastic shield over my nipple. Juan would fill a small syringe with formula, attach a filament-like tube to the tip, then feed the tube under the shield. When Little T would latch on and start feeding, Juan would slowly depress the plunger to supply formula into Little T's mouth.

That was the idea at least. Little T's mouth was so tiny, it was hard for him to latch. That afternoon, after finally latching on, he sucked a few times then defiantly detached, squeezing my nipple between his gums and crying out in protest, a cry that sounded like a really pissed-off duck. I also let out a cry from the pain of him gumming my nipple and from the conviction that this circus act was never going to fly.

This was the sixth time that our attempts at supplemented breastfeeding had failed. Before I was frustrated, but now I was despondent. The stakes were high. Little T had been born small, four pounds ten ounces. We had to get this feeding thing down and his weight up if we wanted to keep him out of neonatal intensive care.

Juan rushed to fill a bottle with formula while I looked down at Little T, as if observing him from very far away. I saw his little limbs kick and thrash, his face turn red as he cried. Somewhere in my brain, I knew I should comfort him. And yet I couldn't move. All I could do was hold my arms in a cradle position so he wouldn't tumble off my lap.

After several more minutes of agitated crying, I stopped caring if he fell. My gaze moved from his little body in its stupid onesie to where my attention really was—at some undetermined point in space where there was a void of feeling and no risk of anger or failure.

Juan came back, saw me staring off into space and Little T slipping from my lap, and ran to gather him up in his arms. He rocked Little T and stuck the bottle in his mouth. Little T sucked voraciously away, instantly quiet, and downed the bottle of formula in under a minute.

That sight brought me out of dissociation, and I began to cry. Breastfeeding was one way I longed to rupture bad cycles. I desperately wanted to give him breast milk antibodies to support his immune system, to give him a better chance of avoiding the sickliness I experienced as a child.

I was disappointed in myself. But I was also angry at Little T. My arms and neck and back all ached from holding him to feed. The stitches from my episiotomy were sore from sitting on them, and they itched uncomfortably as they healed. I blamed him for all of this. Why couldn't he be more patient? Why did he have to be so demanding? Why was he already so spoiled? As I allowed this burning anger to grow, I began to feel wary of myself. My anger was unreasonable and outsize, and I felt capable of doing harm.

At times like these, when the voices of C-PTSD and chronic pain took over my mind and my nervous system defaulted to defense, I had to take breaks from being with Little T. I could feel my shoulders hiked up to my ears, my hands pressed into fists, my back curved like a cat on the prowl. My heart would race and my breath became dammed, and I knew I had to reregulate my own system before I could return to caring for him. Juan knew this too since we had extensive conversations about it when I was pregnant. So whenever I shut down or amped up, which Juan could clearly see from my body cues, he would take Little T to a different room to give me space to re-center. I would take deep breaths, which can act like a neural brake on the sympathetic nervous system. As

I slowly inhaled and exhaled, I would feel my fists relax into open palms and my shoulder blades return to their sheaths.

Parenting with unresolved trauma has been like this: living a split life of knowing my limits and then getting pushed past them by the needs of a newborn child. I feel things I don't want to feel, like anger and resentment toward Little T, and sometimes do things I don't want to do, like yell, scream, and completely shut down. I watch these things happen, and I feel powerless to stop them until I remember that even our most automatic neural responses can be influenced by simple interventions—like breathing, singing, chanting, and feeling supported by others.

So much about being a new parent was triggering to me because so much reminded me of what I had rarely received as a child: eye contact, soothing touch, tender attention, and the patience to do it over and over again. When I was exhausted, which was often during Little T's early years, I often felt that this type of connected presence was out of my league. I needed serving spoons when all I had to give were teaspoons; I needed adult emotional intelligence when my EQ was stuck at age eight. If good-enough parenting were a fat, cakey glazed donut, I sometimes felt like the gaping hole.

I suspected I would face these emotions after birth, and so before Little T was born, I worked to put a number of supports into place. I recruited a postpartum team to help me be a good-enough parent, if not my best self. Juan's mother, Chiqui, agreed to come help us for two weeks. Juan decided to stay home for three months before returning to work, a privilege made possible by our savings but guaranteed to fathers by federal policy in many other countries. I also asked two of my dearest friends—Lisa, who was my doula, and Mia, who was one of my first friends in the Bay Area Filipino activist community—if they could come help us for any amount of time in the month after birth, and they agreed. Not only that, they offered what is virtually unthinkable in our individualistic, capitalist society: they each came for a whole week, using precious vacation time and leaving their own small children to help our little threesome become a family.

~

In my first postpartum month, we enjoyed the stability of a village that passed Little T between many hands. Lisa and Chiqui, my mother and father and brothers, and Little T's cousins all came often to spend time with him.

My mother encouraged Juan and I to take walks while she stayed with Little T. When we came back, we would find my father slouched on the sofa with Little T perched on his chest, both of their mouths wide open, napping. Sometimes we would find my dad holding Little T in his lap while he played the piano. I was happily surprised by how present my parents were. They were tender and attentive grandparents. This was an important part of rupturing the cycle, witnessing them form an attachment with him that they had not been able to form with me.

As I set about trying to develop my own attachment with Little T, I was lucky to have devoted care to support my recovery. I think about the giant sanitary napkins that all birthing people have to use to soak up lochia, the normal postpartum bleeding that lasts for a month or even more. In the first week, my lochia seemed to come from a bottomless spring. Lisa or Chiqui or Juan emptied the garbage whenever it piled up with soaked pads. I think about the ice packs I'd stick in my underwear to ease the swelling where I was stitched up from the vaginal tear. They washed the ice packs and returned them to the freezer so they would be cold when I needed them again. I think about the MiraLAX and senna I ingested to clear painful constipation. They brought me water and tea to wash the medicine down. And then there was my manic emotional state, where I'd feel jubilant one moment and then listless and miserable the next. They led me through breathing exercises and reminded me to close my eyes and rest whenever Little T fell asleep.

If we as a society talked more about the nitty-gritty of recovering from what Angela Garbes calls "the significant medical event" of birth, I believe the village of postpartum support I experienced could become the norm for all birthing people.

Before Lisa returned to California, she cooked a buffet of meals and left them in the freezer for us, including her delicious nilaga, a highly nutritious beef soup. After Lisa left, Chiqui stayed to help us with everything from laundry to cleaning to changing diapers, all the while entertaining us with her lively stories and jokes.

Hours blended into days and days morphed into nights. I got through nightly pumping and breastfeeding and diaper changing by texting two friends from childbirth class for support. Feliza and Ruby's texts always made me laugh. "Who else has boobs hanging down to their belly button?" Feliza texted. "Me," replied Ruby. "Not me (wink emoji)," I replied back. We'd bond over how comical it was to feel delirious and drained and elated and disgusting and sensual and spent and hyperactive and powerful and weak all at the same time.

This small village of care was our version of social childbirth, a phenomenon common in the Philippines as well as many other cultures throughout the world. This village of care calmed my C-PTSD and chronic pain in the first month and buffered me from the worst impacts of sleeplessness and physical recovery.

Yet even with all this support, I was exhausted. Breastfeeding was still a struggle, and it continued to be until Mia came to help. Mia did the incredible service of allowing me to stay in bed for the twenty-four-hour cure. True to its name, the cure calls for twenty-four hours of continuous care for the lactating parent and the baby. During the day, Mia brought meals and washed bottles and the breast pump so Little T and I could stay in bed and focus on feeding at will. It turned out that Little T was a very hungry creature; he would search for my nipple and try to eat at least once every hour. We had been on a feeding schedule of once every three hours, and so he must have been starving by the time I tried to breastfeed him, which explained his agitation and impatience at my slow milk flow. He was a grazer. Just like me. I might never otherwise have known.

Over the next two months, Little T steadily breastfed and gained weight, though we still occasionally had to supplement with formula. In that watery netherworld when day blended into night, I began to cherish the moments when Little T and I would have skin-to-skin time. I would lie down on the bed and rest him on top of my chest, his belly down and head turned to one side, his fuzzy cheek pressed against my breastbone. His whole little body would rise and fall with my breath, and I could feel our heartbeats align. When we had skin-to-skin time, I felt a sense of full-body peace that I had never known before. I had been so focused on what I would or wouldn't be able to offer him that I didn't expect his little nervous system would also have the power to help regulate mine.

~

After Juan went back to work, I stayed home with Little T for six more months, and during this time, I practiced how to ration my spoons and manage my own health while continuing to care for him. I am sure that without this time to work on my parenting, the phantoms of C-PTSD would have severely threatened my ability to continue a bonded relationship with my child.

One afternoon when Little T was five months old, he cried for an hour and a half without stopping. Ninety minutes of uninterrupted infant crying is like ninety minutes of fingernails on a chalkboard. This irritation serves a purpose, to an extent. Our nervous systems change to become more sensitive to infant cries within three months after birthing a child. The parts of our brains necessary to orient, mobilize, and vocalize all become quickly activated, as do emotional processing centers that cause us to react to the cries with a full spectrum of feelings.

I definitely experienced the full range of feels when Little T wouldn't stop bawling. At first, I felt tender and nurturing. I sang, shushed, and rocked him; shook rattles at him; turned on the TV; gave him a sponge bath, to no avail. Then I became concerned. I looked up "what to do when your baby won't stop crying" and followed all the directions I could find. I took his temperature—no fever. I changed his diaper—again. I tried to give him a gentle foot-and-hand massage, with him kicking and screaming all the while and me dodging his little feet and fists. I bundled him up and held him tight and tried dancing and rocking again. And then, I hit a wall. Suddenly, all my spoons were spent, and I had nothing left to give. I felt helpless, overstimulated, and overtaken by a primitive full-body sadness. It was all I could do to walk across the room and deposit him in his rocker before retreating to the couch, where I proceeded to curl up in a little ball and rock myself like a child.

At moments like these, my soul wound felt as if it would swallow me whole. I didn't just feel tired, I felt empty. I longed for a wellspring of nurturing to water my being, to rescue me from the chronic drought of my own childhood neglect. Little T continued to cry, but I was numb to his distress. My ability to empathize with him was obscured by my autonomic nervous system overwhelm. How could I possibly comfort him when I needed someone to comfort me?

Time was my helper and my nurse. I had nowhere to be, no one to report to, no one else to be responsible for but Little T and me. I looked over to make sure he was safe in his rocker. Then, I took ten minutes to myself in the next room with the door closed. I put on headphones and listened to a meditation and took long, deep breaths. I applied a TENS unit to my aching arms to interrupt nerve signals and to provide temporary relief to the pain. When I came back out, I was in a calmer state, a bit more resourced and present, though still weepy and sore. Little T continued to cry, but for the remaining fifteen minutes of this spell, I was able to sit with him in patient company, tilting the rocker back and forth, singing softly and humming, waiting for his cries to pass.

Given the challenges I faced even without having to work a paying job, it is unthinkable to me that our country refuses to prioritize paid leave for all families. In more than a hundred countries across the world, both mothers and fathers have paid family leave of at least three months. In four of these countries, LGBTQIA+ partners of any gender enjoy the same parental leave benefits as heterosexual couples. And in Japan, parents can take up to a year off and still receive some pay the entire time. My experience underscored what I already believed: that all new parents, but especially those with disabilities and mental health challenges, should have paid leave as a matter of public health. It is the only humane way to continue self-care while also raising children through the sensitive neural period of infancy, which can set the courses of their health for the rest of their lifetimes.

During my ten months of parental leave, I was able to learn some important lessons about parenting with C-PTSD. I learned that there is a fine line between staying open to emotions that help parent-child bonding and putting a boundary around emotions that can threaten this bond. Quite simply, emotions are not always your friends. The dance of parenting with C-PTSD is having enough spoons to determine whether your emotional reaction is appropriate to the conditions at any given time, while also resisting primal defense reactions that can be triggered by infants' cries.

At night while pumping, I searched the internet for insight into how to handle this tightrope existence. I read baby boards that reassured me the emotional roller coasters were normal, but I didn't learn much about this particular challenge I was having of parenting while healing attachment-related C-PTSD. What ended up helping me most was gaining a deeper understanding of the neuroscience behind infant development.

Functionally (though not strictly neurologically), our brains develop from the bottom-up. Infants first learn to process sensory information in the lower part of the brain—specifically the brain stem and diencephalon—to regulate their internal bodily processes and emotions. This is why rocking, singing, and skin-to-skin contact are so important. These actions create sensations that are the infant's main vehicle of learning. In a similar way, crying may be an action that allows for much-needed release, not just for emotions but also for the flood of sensory stimuli that can overwhelm babies' bodies and brains. In other words, babies may need to cry simply because it's cathartic.

Learning this changed my reactions to Little T's crying spells. I would watch his body shake with sobs, eyes pinched together and skin turned red, and instead of drowning in my own frustration and sense of failure, I would encourage him to let it all out. Seeing his weeping through the lens of bodily catharsis allowed me to not take his crying personally and to activate neural circuits of care and nurturing while keeping the brake on my own agitation. Suddenly, I didn't feel as defeated and exhausted. My baby was doing what he needed to do.

Of course, frequent crying can also be a sign of medical issues. A month or so later, Juan and I learned that Little T had GERD, a painful condition of acid reflux that is common in low birth weight infants and that persists until the esophagus becomes more fully developed. His doctor prescribed him medication, and he cried a bit less but continued to be colicky. These frequent bouts of incessant crying were the first signs of what we would come to understand as Little T's own neurodivergence. He not only had GERD; he also seemed to have a highly sensitive and atypical nervous system—perhaps inherited epigenetically from me.

Herein lies one of the benefits of parenting with C-PTSD. For those of us with neurodivergent children, we may be able to bring more compassion and a healthy sort of vigilance to raising them. As Little T grew older and I saw how easily he startled, how hard it was for him to calm his body down, how sensitive he was to becoming overtired, and how explosively angry he could become, I was able to treat these reactions with compassion because I experienced them myself. And because I had lived the consequences of these reactions, I knew to take his state of chronic agitation seriously. I understood perhaps better than some neurotypical parents might that he was struggling with different wiring and a baseline of nervous system dysregulation. While some of our friends either catastrophized or minimized his behavior, I knew that these symptoms called for a middle ground of deliberate attention; they, like he, deserved to be seen and held and addressed with supportive services and love.

~

Somatic therapist Resmaa Menakem says, "When we heal and make more room for growth in our nervous systems, we have a better chance of spreading our emotional health to our descendants." This healing, I have discovered the hard way, is ongoing and never linear.

When Little T was three and a half years old, he had a particularly terrible tantrum. He was overtired and overheated and didn't want to take a bath. I closed the door to the bathroom and turned on the faucet, and he became angry and volatile. He kicked and punched and launched his little body at me with a force strong enough to knock me down. And then he proceeded to open up bathroom drawers one by one, launching the contents toward me with all his might: toothbrushes, toothpaste, a nail clipper, hair scissors. I felt a back draft of blistering energy flare up inside my body. My arms extended, and the next thing I knew I had pushed him—hard—so hard he fell backward and hit his head on the floor. He started to cry, and I was satisfied. My nerves were on fire, my eyes clouded by smoke.

And then the smoke cleared. I saw Little T lying on the ground crying, and a tidal wave of remorse washed over me. I gathered him into a tender hug. His crying reduced to a sniffle as he hugged me back with his little arms and wiped his nose on my shoulder. I also began to cry. "I'm sorry," I told him. "I'm so sorry." I asked him if he was okay. He nodded through his tears. I rubbed his head and checked for bruises. There were thankfully none.

Whenever Little T became this physical, my HPA axis, which controls the stress response, would light on fire. My hypothalamus would send signals to my pituitary gland, which would in turn send signals to my adrenals, altogether flooding my body with stress hormones. This in turn made my heart race, my muscles tight, and my cognition limited. Because of my own bottom-up brain processing, my reactions to his tantrums were faster than my thinking brain's ability to put his distress into larger context. Instead of realizing, "Little T is just three, he is upset, this is his only way to express his emotions," my motor muscles went straight into action. The amount of adrenaline that coursed through my body might have been enough for me to climb a three-story ladder and save a child from a burning building. And yet I was mobilized for defense, not for rescue, and so I was in no state to calm him—or myself—down.

Years of therapy had retrained my brain enough to bring me back to my senses before I did anything I would regret even more. My body crashed after the adrenaline rush, and Little T and I sat together on the bathroom floor, sniffling and rocking and hugging, for what felt like an eternity. I was sick with remorse. I decided right there and then that we both needed more help—him for his dysregulated nervous system and the irritation that caused such explosive tantrums, and me for my dysregulated nervous system and the irritation that caused me to respond with outsize anger and aggression rather than compassion and care.

I went back to therapy, this time pursuing somatic touch and EMDR. My new therapist explained that trauma is like the flame that boils water. Trauma symptoms—anger, bodily pain, anxiety, depression—are the bubbles that rise to the top when the water boils. Many modalities treat the bubbles in the boiling water. EMDR treats the flame.

Meanwhile, I started the grueling journey of trying to get help for Little T. Juan and I had no idea where to begin. I read every parenting book I could get my hands on and learned approaches that focused mostly on child behavior. Did we simply need to set better limits? Was Little T just testing us? That was certainly the message we got from some books and from most of the people around us. The side-eyes on the bus when he had meltdowns. The racialized and passive aggressive avoidance from parents who saw him as dangerous. The declaration from a relative that we couldn't let him "get away with acting out" and just had to "nip it in the bud."

What these people never saw is how terrible Little T always felt after "acting out." He often cried afterward and said, "I'm so bad, I can't stop, I can't control myself." My heart ached to see him struggle with anger and frustration that was too much for his small body to hold.

We went through two years of trial-and-error learning: we took parenting classes and worked with a play therapist, an occupational therapist, and a parent-child interaction therapist. Each of these things helped to a degree, but in the end, the interventions that helped most were the ones that focused on his physiological state, not on his so-called conscious behavior. As psychologist Ross Greene says, kids do well when they can. But sometimes, biology gets in the way.

Like me, and like so many of us, Little T was struggling to manage a hypersensitive nervous system, one that was easily overstimulated by sound, touch, and the flood of other sensory inputs from everyday life. We learned occupational therapy exercises to help calm his nervous system and pursued a course of music therapy designed to stimulate his parasympathetic system through a fiber of the vagus nerve in the inner ear. I learned about this form of music therapy when I interviewed Stephen Porges for this book, and I will be forever grateful for this example of how the project of writing can indeed heal.

The results of the music therapy were like night and day. After a week of listening to the program's frequency-altered music, Little T's baseline was much calmer. He was less frustrated by small challenges, less startled by loud sounds, and more able to let his beautifully strong and vibrant personality shine through.

~

There's not much guidance in our society for how to be pregnant, give birth, and parent in the context of trauma. For now, disrupting the cycle of transgenerational trauma is like the process that Juan and I went through to conceive and birth a child—collective and improvisational labor.

My nervous system will always be at one-child capacity. I have always known I would not be able to handle more, and I am eternally grateful that I have had the right and the privilege to plan my family to fit this capacity.

But I am saddened by the many people with wombs who have been denied this right, who have been subject to forced sterilization and forced pregnancies, even before the Supreme Court overturned Roe v. Wade. I shudder to think of how many more will endure the collective trauma of criminalized abortion. It's even more reason that as we work to restore the right to privacy and bodily autonomy, that we also work to build the structures and policies that can prevent enduring traumas in the first place.

The daily challenges of parenting with C-PTSD within an unjust society can send me into inconsolable despair. But every day I am practicing toward nurturing cycles that allow for sources of change.

I watch Little T become one of these sources. He is now eight years old, and though he no longer needs a womb-like environment, he still needs to come snuggle at night. He is healing me while I am raising him, replacing patterns of neglect with experiences of tenderness and connection. And he is already creating new cycles of care.

In this moment when summer has come blazing to Seattle, Little T is learning about rain catchment and the needs of the azaleas and hortensia emerging from late chill in our backyard. He fills his elephant-shaped watering can and lumbers over to the plants. Careful not to spill precious drops, he bends over to touch star roses where their leaves have begun to curl from sudden heat. He tips the elephant toward the plants, and its trunk releases rivers of rain to quench their thirst. I watch him, and I feel my soul wound heal.

What an incredible place to be. The cycle of care is growing. I pledge to continue to nurture Little T like he is caring for these plants, dendritic roots grasping the earth, electric blooms turned toward the sun.

Watershed

this conversation is about us.
the mostly water.
how we will surge,
recede and rise.
—ARLENE BIALA

It is exactly thirty years since my first trip back to the Philippines, and Juan and I are preparing to bring Little T, who is five years old, to the islands for the first time. I feel excited but nervous. Little T feels the same. He is over the moon about flying but worries the plane will run out of fuel. He is excited to meet his relatives in the homeland but is sad about leaving his stuffed-animal family alone for weeks. He looks forward to the beach but is nervous about getting lost.

To calm our jitters, I take him to West Seattle to sit by the Sound. At Alki Point, the water ripples until a white-crested wave laps our shoes in the wake of a passing barge.

"Did you know this water is part of the same larger body of water that reaches all the way to where we're going in the Philippines?" I say.

"Nanay, you've told me that a thousand times," he replies, rolling his eyes.

"Okay, fine, but can you actually imagine it or feel it more when we're close to it?" I ask.

"Not really," he says.

"Close your eyes and listen," I suggest.

He closes his eyes, fidgets, and opens them a few seconds later.

"What do you think?" I ask him.

"I don't know. I'm still nervous," he says. "And I'm hungry."

"Yeah, well, me too."

"That's cuz you're always hungry," he says.

"That's true."

We leave the beach to go find food. What I now wish I'd said to Little T instead: *It makes sense to be nervous; it's a sign that we are open to the imperfections of the world.*

~

Like tributaries reaching for the sea, Juan and Little T and I journey from Seattle to Manila with an overnight stop in Seoul in between. We fly over a vast expanse of ocean, and Little T marvels at what he calls "so much water in the world." For hours after his comment, Bob Marley's "So Much Trouble In The World" plays in my head with Little T's lyrics instead.

Three days after leaving Seattle, we arrive at the Manila Airport, and a live band greets us with "We Wish You a Merry Christmas" as we walk toward baggage claim. It's crowded; just as Little T marveled at the ocean from ten thousand feet in the air, I am in awe of the way people on the ground move like river currents, shifting together when traveling the same way, adjusting to avoid collision when changing direction. I've seen Manila traffic move in a similar way.

We collect our bags and head to the curb. The scent of humidity, diesel, and a tinge of urine reminds me of the last trip I took home with my parents. That was ten years before, a very different time. Back then, my uncle and aunt were still alive. Now my cousins are the elders. Their kids are all adults, and the oldest one, Andrew, has come to pick us up. I see him across the street standing behind a metal barricade, and I wave excitedly while he runs over to help us with our bags. We exchange warm hugs, and I tell him I can't believe it's been ten years.

"I know!" he says. "And now look, you have a husband and son."

"Yeah, and you have a wife and *two* kids," I reply. "You're not just my little nephew anymore."

"You're right, look how much I've grown!" he jokes, widening his arms around his belly.

We all laugh, and Little T pats his cousin's tummy; it seems the banter has made him feel at home.

By the time we cross the parking lot and climb into Andrew's car, Little T is already sweating like a hothouse tomato. Andrew turns the air conditioning on high and promises to drive as fast as he can to cool our Seattle skin. Surprisingly, the traffic is light, and we coast up the highway with ease. I close my eyes and feel my body carried on a concrete axon, moving like an impulse from a peripheral nerve to the brain, to the hippocampus perhaps, or to other locales where long-term memories are stored.

Lauret Savoy writes that to re-member is to find patterns from fragments. From previous trips to the Philippines, I remember a pattern of water: the drenching release of monsoon rains, the rivulets of runoff pouring into the Pasig River, the smell of salt and bumang araw steaming from Manila Bay. I remember fragments of taste and smell: the apple-tart of mangoes from my grandfather's trees and the creaminess of pastillas from Malolos, the faint perfume of my lola's orchids and the mold that drifted from her cupboards. Thirty years ago, I walked into my grandmother's house and saw photos of my mother when she was my age. I realized at that moment that there was a larger pattern to my existence, that our family's story was deep and wide and sourced all the way across the sea.

Andrew pulls into the driveway at my lola's house, which hasn't changed much in three decades. It is a renovated version of the structure my grandfather built nearly a century before, the same house occupied by Japanese officers during World War II. On the mint-green lanai that stretched before us, my lola told thirteen-year-old me her stories about my guerrilla grandfather. Lolo Juan was a lieutenant colonel in the Bulacan Military Area, a prisoner of war who never surrendered information, even under pressure of torture by the Japanese Army. When the war ended, his body was never found.

My grandfather's silence endured for generations. Silence can be honorable when telling would betray the lives of others. Silence can propel us forward when coping is the only medicine. Silence can also take on a life of its own. What fails to come out in words becomes buried in bodies of water—human bodies, riverine bodies—until the words are ready to spring from the source and flow free.

I have considered telling Little T the story of his great-grandfather's heroism but have decided to keep silent for a while longer. He knows that Lolo Juan fought as a rebel during the war, but I will spare him the details till a more age-appropriate time. In the meantime, I try to share stories in other ways.

Inside the living room, I steer Little T to the family photos on the wall. My cousin Annie lives here now with her family, and she has preserved much of the ambience of the old house and many of my grandmother's keepsakes. I point out a picture of Lola Simeona, the namesake for Little T's second name, as well as a smiling portrait of my mother as a teenager. Little T is only passingly interested; when I tell him that's my mom as a girl and that this is the house where she grew up, his mouth drops open momentarily. Then he skitters off to look at the tiny winter village my ate Annie has created on the side table.

Before Andrew drives back to his own home in Manila, he and I plan a ferry ride on the Pasig River the next day. The Pasig is the wandering nerve of the city, like the Chico River is to the Cordillera mountains in the north. As a main channel of the Pasig-Laguna watershed, it enervates dozens of municipalities around the metropolitan area. I've explained to Little T that the Pasig River is important to our family, that it is the source of our people, the Tagalog people, and also the river that my father grew up playing in, back when the waters ran clear. What I want to explain to him now is that the Pasig is also a symbol of the accumulated hardships of Philippine history, a re-membrance of the patterns of all that Filipinos like my grandparents have lived and lost over generations. Since Little T is obsessed with all vehicles, once he hears the word "ferry," he vehemently agrees.

Andrew arrives the next day with the rest of his family. Our plan is to board the Pasig River ferry at Valenzuela and then head west toward Manila Bay. I have a lecture planned for Little T to explain the history of the river and some of the sights along the way. But plans don't always work out in the Philippines—at least not in the way that Americans are trained to believe that plans, once made, are entitled to their execution. Here, life doesn't move in such perfectly calculated and engineered ways. Instead, it moves with the unpredictability of water; the best-laid plans find ways to morph and transform in order to flow around unanticipated obstacles.

We arrive at the Valenzuela terminal and find it closed for repairs. Andrew's wife, Miracle, talks to the security guard, who suggests we drive farther east to the Hulo station instead. We climb back in the car feeling wired and restless, like nerves ready to fire that have not yet reached their threshold. Almost two hours later, after several snaking detours, with the afternoon sun beating through the windows of the car, we finally make it to Hulo, only to find that the last ferry has just left. We have literally missed the boat.

I feel a rush of anger. This was one of my main missions in Manila, and we'd failed. We had wasted the whole day driving for nothing. My American entitlement spills over. My blood boils and I clench my hands into fists. Meanwhile, Little T and his cousins are running around the station playing tag.

We gather them to break the bad news: there won't be a ferry ride after all. Their faces fall, and Little T lets out an angry yelp that makes some people look around with concern.

I call to mind the moment before the trip when Little T and I went to Alki Point to calm our fears. Even then I knew the journey wouldn't be perfect. I had been at peace with this at the time, but now I felt more like Little T and his cousins looked: pouty and disappointed.

Just then the station attendant calls out to us. "Ma'am sir!" We all turn our heads toward this familiar greeting. "There is a ferry coming from Lambingan now," he says. "It is only going one stop to Guadalupe, but maybe ang mga bata will like even a short ride."

Little T looks at his cousins with wide eyes and then up at me and Juan. Miracle and Andrew start nodding, and Juan makes a face that seems to say, "Why not?" and so I exclaim, "Let's do it!" Little T and his cousins cheer, and all four of us jump up and down. Taking a short journey on the Pasig, even if going a completely different direction than planned, is better than no journey at all. Little T continues to bounce like a pogo stick while his cousins twirl in circles. I smile at the adults with satisfaction. Miracle looks at us like she knew this would happen all along.

~

When I conceived of this family trip on the Pasig, I intended to show Little T an example of the impacts of colonization, corruption, and entrenched poverty on our bodies of water. Within the larger Pasig-Laguna watershed, the Pasig is the waterway that has borne the brunt of Manila's history. I recalled its stench from when I had visited at thirteen, when my father took us to see his brother in Tondo and we passed the Pasig along the way. I read about its history as the most polluted river in the world. I wanted Little T to understand the patterns that had broken this storied channel which once upon a time had been a source of life.

In the early twentieth century, after Americans took over colonial rule from the Spaniards, they seeded factories along the Pasig, using the river's waters as a receptacle for industrial waste. Along with waste, bodies also piled up in the riverbed, forming a substrate that made the river at once a liquid mausoleum and eventually a corpse of its own. Bodies from centuries-old conflict with Spanish colonizers, bodies from battles with Japanese and American imperialists, bodies from the current-day war on drugs, all accumulated like silt in the river's depths. During the Philippine-American War, American soldiers stacked the bodies of Filipino soldiers on the banks of this river like ballast.

Because of this accumulated trauma, in the 1990s, biologists pronounced the Pasig River dead. Generations of pollution had choked off all semblance of marine life. The river became a color that made it impossible to see one's reflection, a color that swallowed all hues.

So I brought masks for the kids, as well as extra hand sanitizer and hand towels in case we had to wipe sludge from the seats or disinfect a scrape. I thought we would step onto a rickety platform to board a ferry with cobwebs and cracks in its floor. I imagined the river water would pool inches deep inside.

The reality we find is altogether different. We step onto the floating metal dock behind the station to wait for the approaching ferry, and the first thing I notice is the smell of fresh air. I see that the river is a neutral green-gray, and that the river surface is continuous water unbroken by any signs of trash. The embankments are cement retaining walls painted lively colors, and in some parts, there are concrete steps leading from the barrios down to walkways that line the river's edge. These walkways are alive with people strolling, workers pushing wheelbarrows, chickens strutting, and a few kids wobbling around on bikes. River rocks fortify the embankments, and pockets of water hyacinths float by. The current moves steadily westward toward Manila Bay.

"Wow, this is nice!" exclaims Little T's cousin, and we all agree.

From our vantage point on the ferry dock, the Pasig River does not look dead. Neither does it appear vibrantly alive. Perhaps it is now a revenant, an animated corpse in the process of undoing the effects of death. It is stiff but moving, full of resting potential, flowing with determination toward its next life.

~

The ferry that approaches is a modern boat, white with blue trim, called the Mutya ng Pasig, the Muse of the Pasig. We board and find the boat to be spotless. I feel a bit foolish and stuff the masks and hand towels deep inside my bag. We find seats by two windows, sticking our heads out like dogs in a moving car.

"Now *this* is nice!" says Little T, sweeping his arm to display the length of the boat and then the view outside.

The ferry pulls away from the station, and we are off. Finally we are riding on the river, heading east toward the Guadalupe station, ploughing through the gray-green water and leaving white foam in our wake. River drops spray on my cheek, carrying a faint whiff of something sour before the breeze clears it away. I feel the rumble of the ferry in my belly and the exhilaration of momentum on my skin.

Little T and his cousins clap and squeal. They are having fun, and I am having a moment. I am struck by how different the Pasig River is now compared to when I visited as a child. It seems to be going through a period of traumatic growth, growth that depends on the changeability of water, just as healing depends on the plasticity of nerves.

Traumatic growth also depends on the will of those who would shape the direction of change.

I think about those who reenvisioned the Pasig as a reclaimed waterway. I imagine that such a noticeable transformation required many different bodies coming together to stabilize these troubled waters.

One of these bodies is the Pasig River Rehabilitation Commission, which was formed in 1999, nearly a decade after the Pasig was declared dead. The commission's purpose was to act as a sort of central nervous system, to coordinate the many agencies necessary to resurrect the vital waterway. The commission oversaw the removal of solid waste from the channel and regulated factories to stop dumping into the river. But one of its most successful projects involved a radical departure from previous approaches: instead of only trying to clear the polluted water inside the Pasig, the project also focused on changing the outside environment that contributed to pollution in the first place, one tributary at a time.

The Pasig River has forty-seven minor tributaries, or esteros, that stretch throughout the Manila area, like dendritic trees spread from a nerve cell. In 2010, the commission chose to apply a comprehensive cleanup approach to one tributary called the Estero del Paco. Pictures of before and after the project are startling. Before the cleanup, the estero looked like solid ground—a dumpsite littered with multicolored scraps of plastic, cardboard, and paper. After the cleanup, the estero looks like sweet water—a fluid tributary of deep bottle green lined with landscaped plants and a winding walkway.

Trash from the esteros was a major source of pollution affecting the Pasig's main channel, which we are riding on now. As the ferry hums along, I think about what a revelation it is to be carried on Manila's wandering nerve. Because of the estero cleanup and other coordinated interventions, there are no tires, shoes, or even plastic bags floating in the channel. In fact, as far as our naked eyes can see, the largest threats to the fluidity of the river are the remaining clusters of water hyacinth, invasive weeds that fishermen have mostly picked clear.

I had come prepared to teach Little T about the worst of what we inherit, and instead we are all learning about transformation.

If re-membering is assembling patterns from fragments, then transformation may be the work of breaking old habits to see what new patterns might emerge.

The Pasig has its own story to tell. I am struck by the parallels between the new patterns in the Pasig River and the potential of our bodies to transform the impacts of trauma. I look at Little T and his cousins, who are now kneeling on their seats and sticking their heads and hands out of the window, and remember that when it comes to carrying trauma, it's not that we are polluted but that we bear burdens from environments past and present, burdens that are sometimes so thick, they can keep us from reflecting the sky.

As June Jordan writes, "Neither tiger lilies nor children, by their nature, threaten the rain, or the bees, or the rivers of the world." By nature, we are like the Pasig River, wanting to flow and connect with other bodies of water and to nurture life, which is also, metaphorically, what a healthy nervous system is designed to do. But when we are unwell, as the Pasig River and its entire watershed have been, we must look not just at what is wrong inside our own bodies but at what the external sources of dis-ease might be as well.

The Estero del Paco cleanup took one and a half years and a concerted partnership among government, private corporations, community organizations, schools, engineers, and the Asian Development Bank. Together these parties collaborated to reenvision the entire ecosystem around the estero. They built collective septic tanks. They planted attractive vegetation that further filtered the water. They seeded neighborhood-based recycling programs, dredged the estero, and cleaned drains clogged with trash.

Bantay Kalikasan, a local nonprofit with deep community roots, endeavored to relocate informal settlers in a socially responsible way. They moved families to better and more stable housing, where they were promised access to job training. For those who stayed, they helped train paid "River Warriors." These River Warriors continue to clean trash out of the estero, enforce the city's plastic bag ban, and encourage residents to keep the estero clean.

The Estero del Paco cleanup has become a showcase of how change can begin with a small-scale, but multifaceted, intervention. It is an example of what adrienne maree brown calls the "small actions and connections [that] create complex systems . . . that [then] become ecosystems and societies." The Estero del Paco cleanup was a keystone step toward treating the Pasig in relation to its environment, toward a whole river reclamation as well as the sustainability of the Metro Manila watershed as a whole.

Several years later, the estero still requires continuous monitoring and further cleanup. But it flows more freely than before, the area is more resilient to flooding, and now the estero waters reflect the sky.

~

A few days after the ferry ride, I try to visit the Estero del Paco. But I never quite make it because the area is not exactly safe for outside visitors. I learn this from two organizers with Bantay Kalikasan, who had agreed to take me there, then changed their minds. Instead, the organizers take me to the Estero de Santibañez, which has a similar cleanup story to that of the Estero del Paco. But around the Estero de Santibañez the organized community has established a zone of relative safety, a climate that local leaders work hard to preserve.

One of these leaders is Ate Angie. Ate Angie is a sixty-something-year-old woman with smiling eyes and wiry hair pulled back in a ponytail. She is a River Warrior who earns a stipend of ten thousand pesos a month to steward the estero. Her job includes removing trash from the water and educating community members on how to keep the river clean.

Ate Angie escorts me down the winding pathway that lines the tranquil estero. We pass children playing on a parked motorbike and two yellow plastic bags full of crumpled aluminum cans. She motions to the top of the embankment, which has a wide cement cap that serves as a bench, and invites me to sit down under the shade of a young tree.

Ate Angie tells me—in slow Tagalog so I can understand—that her dedication to the river has been lifelong. She grew up alongside the Estero de Santibañez, and remembers swimming in it, as my dad remembers swimming in the Pasig farther downstream. When the river grew more polluted, children got sick. They developed asthma and stomach problems, and some could not get out of bed. Families could no longer sit by the river because it smelled so bad. They had to stay inside their already cramped quarters, which led to more trouble and fighting within households. So she and other mothers organized themselves to solve the problem.

The afternoon breeze sweeps a wisp of hair across Ate Angie's face. I try to detect a remnant of the bad odor she described, but all I smell is a whiff of detergent coming from behind us, where an elderly woman squats by a plastic tub, washing clothes. Ate Angie looks out on the water as one might look down proudly at a child. "Kasing lusog natin ang ilog," she says. "We are only as healthy as our river."

For years, she and other mothers volunteered their labor, trying to keep the estero clean. They made important progress, but it took the concerted efforts of the Pasig River Rehabilitation Commission to give this grassroots work the resources and multisector support needed to transform the estero from unhealthy to relatively clean.

The children who had been playing on the motorbike are now running in circles. One of them knocks over a yellow bag filled with cans and immediately sets about picking all of them back up.

"I want to show you something," Ate Angie says to me in English. She climbs down the embankment ladder onto a wide plastic raft. "This is my—paano mo sasabihin?—my cruise ship," she says, laughing, "This is what I use to take the garbage out of the water." She proudly shows me the tool she created to do so: half of an electric fan grill attached to a long wooden pole. She uses the guard grill as a net to scoop up a tube of toothpaste, a plastic bottle full of river water, and an empty bag of potato chips, continuing to talk to me while she works.

Ate Angie shares her ten-thousand-peso stipend with the children who help collect recycling. In this way, she distributes both the income from her labor and also the lesson of shared security through collaboration for environmental action.

At the end of our time together, she tells me she is close to losing her stipend. President Duterte had just announced the abrupt dissolution of the commission, claiming the Pasig River could no longer be cleaned. But she says no matter what happens, she will continue to take care of the river and teach others to do the same. I tell her that she is an example of something I've learned about healing, that healing is dependent on a sense of safety and that, as Gabor Maté says, safety is not the absence of threat but the presence of connection.

"You are healing the river and your community," I observe.

"And also myself," she agrees as she stuffs the water bottle in a yellow recycling bag and the tube of toothpaste and chip bag into the trash. "I have clean water in my heart."

~

I watch the white peaks of the ferry's wake settle into the dark river's surface and think about how to adjust my conversation with Little T after the ride. I had been hoping to teach Little T a lesson in history, and here I was learning a lesson in hope. I had come expecting to show him the worst of the impacts of transgenerational trauma on our bodies of water. Instead, I saw an example of the possibilities of transforming trauma not just from the inside out but also from the outside in.

We have left the towers of Makati behind and are now passing the high-rises of Ortigas. This is where many of the Philippines's call centers are located and where some of my cousins are employed. They work long hours, including graveyard shifts, to fulfill the twenty-four-hour customer service needs of foreign consumers and large corporations. When it comes to land use in the Pasig-Laguna watershed, it is the squatter communities who get blamed for throwing trash and sewage into the river, but high-rises and the businesses that occupy them also contribute to the problem. The volume of concrete that is poured to ground these buildings has long replaced the floodplains that would otherwise allow stormwater to return to the earth. Only 5 percent of Metro Manila is connected to a sewage system, and only 10 percent of wastewater is treated at all.

The estero cleanups address some of this land-use problem by creating localized sewage treatment and by restoring greenways with native plants that filter water and help absorb flooding. Though estero cleanups alone are imperfect, and not a solution to wide-scale reclamation, they are a glimpse of what could be if the entire Metro Manila watershed were managed in a sustained, integrated, and coordinated way.

A short twenty minutes after boarding, the ferry hum quiets to a murmur as we slow down under the Guadalupe Bridge. We are nearing our first and final destination. Little T and his cousins haven't noticed that the ride is about to come to an end. They are enjoying themselves thoroughly, pointing out the window and imagining that the riverbanks are covered in unicorns.

"There's a purple one right there!" squeals Little T's cousin.

"Where?" asks Little T.

"Right there on the stairs!"

"No, I think that's a narwhal," says Little T, laughing.

Their imaginative play is inspiring. They are envisioning the impossible.

When Little T and his cousins realize the ferry is slowing down, they pause their game and follow me to the back of the boat. And then we see it—a shining thing the color of the river, moving as if the water has just taken form. It is a single fish jumping at the side of the ferry. Its scales turn from gray to iridescent black in its brief moment in the sun.

The impossible is happening. The river is resurrecting, and its waters are generating new life.

~

One of my most trauma-wise friends, Spenta Kandawalla, says that the parts of us affected by trauma are dormant, not dead, waiting to be awakened to their potential. In describing her acupuncture practice and somatic healing work, she says, "We don't ask what's wrong with you, what's broken in you. We start with: Where you are going, and what do you long for? These invite the dormant parts to come back to life and answer us."

In our bodies, there are answers. These answers spring from a source called interdependence. The very source of the name "Pasig" is a Sanskrit word that means "river flowing from one body of water to another." The COVID-19 pandemic disabused many of the illusion that our bodies are rigid and self-contained entities. Much like our oceans are connected to rivers and streams, our bodies are connected within larger networks of health and dis-ease. As my friend Malkia Devich-Cyril has said, "We are each other's immune systems." We are also each other's nervous systems.

When we talk about healing trauma, what we really mean is the potential of our interdependent nervous systems to change and grow, and to bring dormant parts of ourselves back to life. The paradox of this plasticity is that negative changes can also become entrenched over time. The entrenched effects of unresolved trauma constitute a public health crisis just as the pollution and floods of a dormant waterway can put entire populations in harm's way.

As a society we deal with unresolved trauma as if we were each independent, as opposed to interdependent. The approach is reactive and does not address traumatic systems at their source. While we can achieve a certain amount of trauma healing through individual, downstream interventions, logic and science show that we can only achieve community-wide health if our larger environments change upstream as well.

This is what it means to become trauma-wise, rather than trauma-responsive or even trauma-informed. When we become trauma-wise, we accept trauma as part of living, but work toward more equitable and coordinated systems for addressing its long-term harms. The work of being trauma-wise requires a vision ample enough to reorganize society toward trauma prevention on a collective scale. As Judith Herman writes, trauma work is fundamentally collective and political since it brings into focus whole populations of people affected by traumatic systems of oppression. As Aurora Levins Morales suggests, trauma healing, at its root, is therefore the work of systemic political and cultural change. This process looks less like traditional medical care that treats disease as problems of individual bodies, and more like the integrated work of watershed management, which takes an ecosystems approach toward the well-being of all living things.

~

By now we've left the ferry terminal, and Little T and his cousins are bouncing up and down the cement steps at the side of the river.

"Did you like the ferry ride?" I ask them.

"Yes!" "Yeah!" "Of course!" they all scream.

"Did it make you think of anything?" I say, fishing for company in this moment of reflection and revelation.

"Like what?" Little T asks.

"Like how connected we are to the water or to the history of this place . . ."

"Uh . . ."

"Look!" Little T's cousin squeals. She points toward a sorbetero wheeling his cart in our direction. What the kids want to connect with is ice cream. The sorbetero offers to make each one of us a cone with all three flavors: coconut, mango, and melon. I smile at the hustle-driven belief that we can and should have it all.

On the way home from the ferry ride, with the kids' bellies full of sorbetes, Andrew drives us back through the metropolitan area, navigating the roads like an embolus journeying through narrowed arteries. Though we complain about the traffic, we are of course part of the problem.

We take a detour to Tondo to give Little T a glimpse of the barrio that my father was from. We drive down Velasquez Street, creeping behind motorcycles carrying two to four people at a time. A tricycle driver pedals with filleted fish drying on the cab's tin roof. Small boys rap on the side windows, waving sampaguita wreaths and boxes of gum for sale.

"Can you imagine Lolo growing up here?" I ask him.

Little T shakes his head, then says, "Well, sort of."

"He was actually one of these boys selling gum, except he sold lottery tickets," I tell him.

I watch Little T stare intently at one of the boys in the street, a kid not much older than him, wearing a long Michael Jordan jersey and tsinelas. Then he sees three children about his age sitting on top of a tricycle cab roof and asks excitedly, "Did Lolo ever do that?"

"Probably not," I reply. "They had horse-drawn carriages back then."

"Why can't I do that?!" Little T exclaims.

Soon we pass a tributary of the Pasig that is so littered with garbage, it is hard to tell there is actually water underneath. I point it out to Little T.

"That's the estero that Lolo used to swim in."

"What?!" Little T shouts. "But it's so dirty!"

"Yeah, it is," I agree. "But it wasn't always that way. Lolo told me when he was your age, it was clean."

I look back at the estero as we drive past, and my heart sinks, weighed down by the difference between this tributary in Tondo and the one I had seen at the Estero de Santibañez. The work of reshaping society to address the roots of trauma is no different from the work of reclaiming water and reshaping our environments toward justice—there may be progress, but there are also setbacks and inequities. We have a long way to go.

~

As we slowly make our way home, I once again think of our car as an electrical impulse on another nerve in the city's body, this one an even more leisurely, unmyelinated nerve than the Pasig. We will get there. We will reach our destination. It requires having patience with our ghost-haunted bodies, enough to board the slow boat that hums through troubled rivers. This boat navigates by the North Star of our longings, toward a delta where sweet and salt waters mix to create imperfect but rapturously interdependent bodies of health.

We are facing a watershed moment where healing from trauma is a generational call. How will we answer it? Will we reproduce old patterns of divide-and-conquer, especially in the face of rising tides of mental health crises and climate change? Or will we re-member and reenvision our interdependence and flow toward collective transformation? The answers, and new questions, are up to us.

In the meantime, I hear callings from Little T; from my lost brother, BJ; from my lola, who broke silence; from my lolo, who kept secrets; from my mother, who grew gardens; from my father's iron will and surgical precision resurrected as an ethereal guide. I feel blanketed by the mist of trauma wisdom, which is the wisdom of the ancestors. I thank my body for bearing their lessons, for carrying the candle of freedom dreams that burn. I thank all nervous people who sense the dangers of the world and still choose to love. I thank all children for giving us the chance to change. I thank water for showing us the way.

Acknowledgments

If it takes a village to raise a child, it took a city to birth this book. First, thank you to my partner Juan for bringing me to a threshold of well-being that allowed me to put pen to page. Thanks to my mother and mother-in-law, who helped care for Little T during MFA residencies and writing retreats, and to Ticha Maleegrai, Robb Valentine, Natalija Colic, Cory Rivard, Lisa Querido, Sylvia Spicer, and Chelsea Summers for pinch-hitting after-school care. Gracias a Leo y Claudia Martell, me ayudaron cuando tenía fechas de entrega y episodios de enfermedad, ustedes no solamente limpian sino cuidan al corazón de la familia y de nuestro hogar.

Nervous grew from a lineage of indigenous, Filipina/o/x, Black, and Latine scholars that includes the work of Karina Walters, Maria Yellow Horse Brave Heart, Eduardo Duran, and Robin Wall Kimmerer; Leny Mendoza Strobel, E. J. R. David, Kevin Nadal, and Joyce Javier; Audre Lorde, Joy DeGruy, Makani Themba, and Saidiya Hartman; and Gloria Anzaldúa and Aurora Levins Morales. Thank you also to Nancy Krieger for her essential framework on ecosocial public health, and to Makani for introducing me to her work. Lorde's *The Cancer Journals* and Morales's *Medicine Stories* were my beacons: more than any other books, these helped me see the possibilities of wielding personal truth to explore the intersections of culture, race, gender, illness, and oppression. Maxine Hong Kingston's *The Woman Warrior*, Grace Talusan's *The Body Papers*, Melissa Febos's *Abandon Me*, Hasanthika Sirisena's *Dark Tourist*, Esmé Weijun Wang's *The Collected Schizophrenias*, and Bhanu Kapil's *Schizophrene* also greatly influenced my writing.

Gratitude to my mentors at the Rainier Writing Workshop. Barrie Jean Borich laced me up on creative nonfiction written by a colorful and queer canon of memoirists, essayists, and hybrid poet-writers. Her guidance on my MFA thesis launched the ship that became this book.

Julie Marie Wade coached me through early versions of "A Brief History of Her Pain" and taught me to cultivate my own intuition about when a piece is done. Kent Meyers modeled the mystery of process as well as the beauty of lyric fiction. Brenda Miller gave me the idea of a vehicular journey for the essay that became "Mobility." Marie Mutsuki Mockett challenged me at a critical time. Rick Barot proved that MFA programs can be diverse and supportive places and encouraged me toward excellence. I am indebted also to the Illinois Mathematics and Science Academy and the History of Science program at Harvard for teaching me to look at science through the lenses of race, class, gender, and justice.

Love to the dear friends who waded through early drafts: Emily Wang, Daniel Colón-Ramos, Sara Gharbi-Reinking, Amy Sonnie, Shilpi Suneja, Mia Villanueva, Lisa Juachon, Martin Gowdy, and Spenta Kandawalla. Your attention helped me believe I could do this, and your wisdom stayed with me through the entire process. Spenta also talked me through a tough moment when I wanted to flip tables. Thank you to Abi Pollokoff for invaluable developmental edits and brownies, and Laura Da' for the soul read. Your feedback was a true act of what Natalie Diaz calls generosity. A world of thanks also to the beta readers who accompanied me on the last stretch of this journey: Adriana Rambay, Lisa Factora-Borchers, Sunu Chandy, Ro Alegría, and Constance Collier-Mercado. Your artistry is embedded in these pages.

Appreciation for all my cowriters on the other side of Zoom: Karina Muñiz, Joyce Chen, Chekwube Danladi, Serena W. Lin, Jhani Randhawa, Leslie Tucker, M. Jacqui Alexander, JoAnn Balingit, Sangu Iyer, Leila Nadir, Lauren Taylor, Tijanna Eaton, Swati Khurana, Ame Gilbert, and Marian Ryan, and to my official and unofficial book coaches: Minal Hajratwala, whose guidance was priceless; Grace Talusan, for being my Filipina memoir mentor; Drea Aguilar, who introduced me to my highest self; Malkia Devich-Cyril, whose emotional and intellectual genius is a primary source of strength in my life; Amy, my ride-or-die thought partner and active voice queen; Angela Garbes, who paved the way and gave great advice; Geeta Kothari, one of the smartest editors around; Anastacia-Renée for the moral support through messaging; Sasha LaPointe and Kristen Millares Young for wisdom at key times; and Manami Kano, for the walks and talks, laughs and political feedback.

Shout-out to the 2019 Jack Jones Fellows, the Hugo House Gush Guts coven, The Unicorn Authors Club, and the Luna Moon Poets Group for being my best touchstones of literary community.

One of my favorite parts of writing this book was connecting with people through interviews. Thank you to the scientists, scholars, and therapists who shared their expertise: Bianca Jones Marlin, Coleen T. Murphy, Peggy Mason, Hongjun Song, Rachel Yehuda, Brian Dias, Stephen Porges, Ate Leny, E. J. R. David, Pat Ogden, Shawn Ginwright, Jeanie Tietjen, Phyllis Thompson, Janice Carello, and Alex Shevrin Venet. Thank you to those who trusted me with your stories: Ate Angie, Alexis David, AD, Imee Dalton, Aimee Espiritu, Cristina Mitra, Leo Esclamado, Janet Stickmon, Hazel Benigno, and Jessy Zapanta. And a special thanks to those who coremembered with me: Lyn Alisago, Roseli Ilano, Paul Bolick, Lisa Juachon, Andrew Mendoza, and also Ruby Linsao for her spirit memory.

Thank you to Jen Santos and Melvic Cabasag for taking me to the Estero de Santibañez and to Ipat Luna for connecting me with them. Love to Feliza Guidero for animal medicine and trifecta TMIs, Bernice Yeung for your life-changing referrals, Roseli for the thoughtful gift of *Concepcion* and the hot springs visioning session that helped spur my writing journey, and Lyn for the sisterhood, the hard but soul-filling conversations, and all the laugh-cry-laughs.

Gratitude for the social justice family who encouraged me to take space to write and who took time to read my writing: Joseph Phelan, Kim Freeman Brown, Raine Brandon, Beulah Osueke, Hermelinda Cortés, Alejandro Cantagallo, Jasmine Leeward, Sam Robles, and especially Belma Gonzalez for your bruja spidey-sense and the gift of envisioning spaciousness. Thank you to Joseph also for the expert comms advice, and to Alice Wong for your inspiring sisterhood, access intimacy, alley-oops, red lipstick, and wonderful conversations.

The Vermont Studio Center, the Jack Jones Literary Arts Retreat, the Dorland Arts Colony, and the Dairy Hollow Writers' Colony provided much-needed space that allowed me to make significant progress on the book, from first proposal and draft to major revisions. Erin Stalcup at *Waxwing*, Melissa Sipin and Bel Poblador at *TAYO*, and Erin Jones and Paul Rabinowitz at ARTS By The People provided homes for previous versions of "A Brief History of Her Pain"; "Loverboy," which I wove into "Mobility"; and "Unbroken Water."

To the fierce women who believed in me and championed my work from jump: my agent, Samantha Shea of Georges Borchardt, and my editors Jennifer Baker and Francesca Walker—I am a better writer because of you, and the publishing industry is a better place because of you. Gratitude to the best copyeditor ever, Kaitlyn San Miguel, and to Abby West, Makayla Tabron, Andrew Jacobs, Stephen Brayda, and the whole design, production, and promotions teams who helped *Nervous* come to life.

To Little T: thank you for your comedy and your cuddles. To my parents and grandparents: thank you for your stories and your silence, your belief in my abilities, your legacies.

Notes

Epigraph

vii *"Historical trauma is a story of love":* Karina Walters, "UW-Social Work-Leading Lights Series-Dr. Karina Walters-2021-02-25-330pm Pacific Time," UW School of Social Work, streamed live on February 25, 2021, YouTube video, https://www.youtube.com/watch?v=dtHZOKLG5xI.

vii *"What we have is a knowing body":* Merlinda Bobis, "Decolonial Poetics: Panel with Merlinda Bobis and Rick Barot," The Digital Sala, streamed live on August 18, 2020, YouTube video, https://www.youtube.com/watch?v=5ao95jpxGjo.

vii *"trying to do the right thing":* Stephen W. Porges, *The Polyvagal Theory: Neurophysiological Foundations of Emotions, Attachment, Communication, and Self-Regulation Theory,* (New York: W. W. Norton & Company, 2011).

Author's Note

xi *the term Filipinx:* For more on the history and politics of the term "Filipinx," see Kay Ulanday Barrett, Karen Buenavista Hanna, and Anang Palomar, "In Defense of the X: Centering Queer, Trans, and Non-Binary Pilipina/x/os, Queer Vernacular, and the Politics of Naming," *Alon: Journal for Filipinx American and Diasporic Studies* 1, no. 2 (2021): 125–47, doi:10.5070/LN41253177.

xi *"its variations":* For a comprehensive discussion of terminology, see Kevin L. Nadal, *Filipino American Psychology*, 2nd ed. (Hoboken: Wiley, 2021).

Introduction

1 *"a story I'm not supposed to tell":* This is an homage to "No Name Woman," the opening essay of Maxine Hong Kinston, *The Woman Warrior* (New York: Alfred A. Knopf, 1976).

2 *Lorde:* Audre Lorde, *The Cancer Journals* (San Francisco: Aunt Lute Books, 1980).

2 *central sensitivity syndrome:* Muhammad B. Yunnus, "Central sensitivity syndromes: a new paradigm and group nosology for fibromyalgia and overlapping conditions, and the related issue of disease versus illness," *Seminars in Arthritis and Rheumatism*, 37, no. 6 (2008), 339–52, doi:10.1016/j.semarthrit.2007.09.003.

2 *peripheral neuralgia:* "Neuralgia (Peripheral Neuralgia)," International Neuromodulation Society, https://www.neuromodulation.com/peripheral-neural gia-definition.

2 *dystonia:* "Dystonias Fact Sheet," National Institute of Neurological Disorders and Stroke, https://www.ninds.nih.gov/dystonias-fact-sheet.

2 *mild scoliosis:* John Philips, "Mild Scoliosis," Southwest Scoliosis and Spine Institute, January 10, 2021. https://scoliosisinstitute.com/mild-scoliosis/.

2 *generalized anxiety disorder:* "Generalized Anxiety Disorder (GAD)," Johns Hopkins Medicine, https://www.hopkinsmedicine.org/health/conditions -and-diseases/generalized-anxiety-disorder.

2 *social anxiety disorder:* "Social anxiety disorder (social phobia)," Mayo Clinic, https://www.mayoclinic.org/diseases-conditions/social-anxiety-disorder /symptoms-causes/syc-20353561.

2 *C-PTSD:* "What is complex PTSD: Symptoms, treatment, and resources to help you cope," Medical News Today, https://www.medicalnewstoday .com/articles/322886. Stephanie Foo wrote an excellent book about her personal experience with C-PTSD. See Stephanie Foo, *What My Bones Know: A Memoir of Healing from Complex Trauma* (New York: Ballantine Books, 2022).

2 *Seven out of ten*: Ronald C. Kessler et al., "Trauma and PTSD in the WHO World Mental Health Surveys," *European Journal of Psychotraumatology* 8, no. 5 (2017), 1353383. doi:10.1080/20008198.2017.1353383.

2 *About one in ten:* "How Common Is PTSD in Adults?," PTSD: National Center for PTSD, https://www.ptsd.va.gov/understand/common/common _adults.asp.

2 *one in five people worldwide:* "PTSD Statistics Worldwide," CFAH, https:// cfah.org/ptsd-statistics/#PTSD_Statistics_Worldwide.

A Brief History of Her Pain

3 *"I will keep Myself":* Alice Walker, "I Will Keep Broken Things," Alice Walker: The Official Website, December 9, 2010, https://alicewalkersgarden .com/2010/12/i-will-keep-broken-things/.

4 *the Kahun Papyrus:* Helena Trindade Lopes and Ronaldo G. Gurgel Pereira, "The Gynaecological Papyrus Kahun," IntechOpen, October 13, 2021, https:// www.intechopen.com/chapters/78710.

4 *dislocated or starved uterus:* John M. Stevens, "Gynaecology from ancient Egypt: The papyrus Kahun: A translation of the oldest treatise on gynaecology that has Survived from the ancient world," *Medical Journal of Australia* 2, no. 25–26 (1975), 949–52, doi:10.5694/j.1326-5377.1975.tb106465.x.

4 *priestesses:* Rhoda Wynn, "Saints and Sinners: Women and the Practice of Medicine Throughout the Ages," *JAMA* 283, no. 5 (2000), 668–69. doi:10.1001 /jama.283.5.668-JMS0202-4-1.

4 *hysterical:* The word "hysterical" has been used to denigrate women's illness

and emotions for centuries now. The roots stem back to ancient Greece, since the Greek word for "uterus" is "hystera." Hippocrates was the first to coin the term. See Cecilia Tasca et al., "Women And Hysteria In The History Of Mental Health," *Clinical Practice and Epidemiology in Mental Health* 8 (2012), 110–19. doi:10.2174/1745017901208010110.

6 *emotionally erratic behavior:* Some believe that the term comes from "hysteria," and that hysterectomies were performed to treat hysteria. But what the two share in common is just the root word "hystera," which is the Greek word for "womb." Although hysteria was attributed to errant behavior of the womb, there is no evidence that hysterectomies were ever used to treat hysteria. Instead, hysterectomies were performed for physical problems in the womb, such as an inverted or gangrenous uterus. See Zouhair Odeh Amarin, "Hysterectomy: Past, Present and Future," IntechOpen, February 27, 2022, https://www.intechopen.com/chapters/80633.

6 *wandering womb:* Morgane Laffont, "Hippocrates and the Concept of the 'Wandering Womb,'" Academia.edu, 2019, https://www.academia.edu/38692894/Hippocrates_and_the_concept_of_the_Wandering_Womb_?auto=download.

6 *tendency to move:* Matt Simon, "Fantastically Wrong: The Theory of the Wandering Wombs That Drove Women to Madness," *Wired*, May 7, 2014, https://www.wired.com/2014/05/fantastically-wrong-wandering-womb/.

7 *sexual healing:* Sander L. Gilman et al., *Hysteria Beyond Freud* (Berkeley: University of California Press, 1993), https://publishing.cdlib.org/ucpressebooks/view?docId=ft0p3003d3;query=bible;brand=ucpress.

9 *cure is punishment: exorcism:* Tasca et al., "Women And Hysteria In The History Of Mental Health," 110–19.

9 *the Devil is to blame:* Arran Birks, "The 'Hammer of Witches': An Earthquake in the Early Witch Craze," *The Historian*, January 24, 2020, https://projects.history.qmul.ac.uk/thehistorian/2020/01/24/the-malleus-maleficarum-an-earthquake-in-the-early-witch-craze/.

9 *gallows, strangling, and beheading:* Ben Panko, "Last Person Executed as a Witch in Europe Gets a Museum," *Smithsonian Magazine*, August 29, 2017, https://www.smithsonianmag.com/smart-news/last-witch-executed-europe-gets-museum-180964633/.

10 *as many as one hundred thousand:* Anne L. Barstow, *Witchcraze: A New History of the European Witch Hunts* (San Francisco: HarperOne, 1995).

10 *Most are healers:* For more on the ways the Christian culture anathematized women and effectively separated the feminine and women themselves from the evolving world of Western science and medicine, see David Noble, *A World Without Women: The Christian Clerical Culture of Western Sciecne* (New York: Knopf, 1992).

10 *"goes about in the garb of a woman":* Tapar created a syncretic religion that was a threat to traditional Catholicism. They are considered to be evidence of homosexuality and gender-bending figures in early colonial Philippines.

See Jean-Paul G. Potet, "The Family," in *Ancient Beliefs and Customs of the Tagalogs* (Morrisville, NC: Lulu Press, 2017), 97.

11 *crocodiles are said to have finished them off:* Henry F. Funtecha, "The Tapar uprising in Oton, Iloilo," *The News Today*, February 9, 2007, http://www.thenewstoday.info/2007/02/09/the.tapar.uprising.in.oton.iloilo.html.

12 *aftereffects of a traumatic event:* M. S. Micale, "Charcot and *les névroses traumatiques*: Scientific and historical reflections," *Journal of the History of the Neurosciences* 4, no. 2 (1995), 101–19, doi:10.1080/09647049509525630.

12 *talking it out:* Pavi Sandhu, "Step Aside, Freud: Josef Breuer Is the True Father of Modern Psychotherapy." *Scientific American*, June 30, 2015, https://blogs.scientificamerican.com/mind-guest-blog/step-aside-freud-josef-breuer-is-the-true-father-of-modern-psychotherapy/.

13 *"chimney sweeping": John Launer,* "Anna O and the 'talking cure'," QJM: An International Journal of Medicine 98 (2005), 6, https://doi.org/10.1093/qjmed/hci068.

13 *"Studies on Hysteria":* Ray Dyer, Josef Breuer and Sigmund Freud's *Studies on Hysteria* [1882] 1893, 1895," The Victorian Web, February 25, 2021, https://victorianweb.org/science/freud/hysteria.html.

13 *Anna E. is a writer:* "Bertha Pappenheim," Wikipedia, https://en.wikipedia.org/w/index.php?title=Bertha_Pappenheim&oldid=1108606259.

14 *Bertha published:* Judith Herman, *Trauma and Recovery: The aftermath of violence—from domestic abuse to political terror* (New York: Basic Books, 1992). This is a classic in the field of trauma healing. Before there was Bessel van der Kolk and *The Body Keeps the Score*, there was Judith Herman and *Trauma and Recovery.*

15 *Hysteria becomes divided:* Carol S. North, "The Classification of Hysteria and Related Disorders: Historical and Phenomenological Considerations," *Behavioral Science* 5, no. 4 (2015), 496–517. doi:10.3390/bs5040496.

16 *chronic pain disorder are myriad*: This article was a turning point for me in learning about chronic pain. It was the first evidence I found that my pain could be due to a physiological condition and not just something I was making up in my own head. See Alice Park, "Healing the Hurt," *TIME*, March 4, 2011, https://time.com/83461/healing-the-hurt/.

16 *the fifth vital sign:* David W. Baker, "The Joint Commission's Pain Standards: Origins and Evolution," The Joint Commission, May 5, 2017, https://www.jointcommission.org/-/media/tjc/documents/resources/pain-management/pain_std_history_web_version_05122017pdf.pdf?db=web&hash=E7D12A5C3BE9DF031F3D8FE0D8509580&hash=E7D12A5C3BE9DF031F3D8FE0D8509580.

16 *the Girl Who Cried Pain:* Diane E. Hoffmann and Anita J. Tarzian, "The Girl Who Cried Pain: A Bias Against Women in the Treatment of Pain," Journal of Law, Medicine & Ethics 29 (2001), 17, doi: 10.1111/j.1748-720x.2001.tb00037.x.

16 *One in five people*: Daniel S. Goldberg and Summer J. McGee, "Pain as a

global public health priority," *BMC Public Health* 11, no. 1 (2011), 770, doi: 10.1186/1471-2458-11-770.

16 *most of them women:* "Pain in Women," International Association for the Study of Pain (IASP), https://www.iasp-pain.org/advocacy/global-year/pain-in-women/.

16 *undertreated for pain:* Carmen R. Green et al., "The Unequal Burden of Pain: Confronting Racial and Ethnic Disparities in Pain," *Pain Medicine* 4, no. 3 (2003). 277–94, doi:10.1046/j.1526-4637.2003.03034.x.

17 *how we transcend it:* Joan Wylie Hall, ed., *Conversations with Audre Lorde* (Jackson, MI: University Press of Mississippi, 2004), 16.

17 *how unbearable it is:* The first edition of *Medicine Stories*, published in 1998, was my first introduction to the type of writing I most wanted to do: personal stories mixed with political perspective. It's an early model for this book, which is my own medicine story.

Aurora Levins Morales, *Medicine Stories: History, Culture and the Politics of Integrity*, rev. ed. (Durham, NC: Duke University Press, 2019).

17 *the absence of everything:* Eula Biss, "The Pain Scale," *Creative Nonfiction*, no. 32 (2007), 65–84.

17 *the pain not be wasted:* Lorde, *The Cancer Journals*.

18 *our mothers' gardens:* Alice Walker's essay "In Search of Our Mothers' Gardens" has long been an influential one for me. Her description of the creative energy of generations of Black women thwarted by slavery and patriarchy yet powerful enough to find expression through acts like gardening, resonated with what I know of my own matriarchal lineage of creativity. The act of writing is sacred to me for many reasons, not the least of which is the feeling of writing the voices of ancestors who were unable to record their own stories because of misogyny and colonization. I am thankful for the privilege of expressing a long line of creativity through writing, of getting to cultivate our mothers' gardens, when my ancestors did not have the conditions to do so.

I: Neurogenesis

19 *in utero through teenage years:* Peggy Mason, *Understanding the Brain: The Neuroscience of Everyday Life*, Coursera class (Chicago: University of Chicago).

19 *"I can believe almost anything":* Michelle Peñaloza, "On Migration, Upon Finding an Old Map" in *Former Possessions of the Spanish Empire* (Riverside, CA: Inlandia Institute, 2019), 58.

Nervous

28 *fueled by conscious choice:* The brain hasn't always been the center of attention; for most of history before the twentieth century, healers and philosophers focused on other parts of the body, such as the heart, as the seat of our most important functions. Byrd Pinkerton, "How technology has inspired

neuroscientists to reimagine the brain," *Vox*, November 17, 2021, https://www.vox.com/unexplainable/2021/11/17/22770720/brain-science-technology-neurology-matthew-cobb.

28 *nineteenth century, Western views:* Benjamin Ehrlich, *The Brain in Search of Itself: Santiago Ramón y Cajal and the Story of the Neuron* (New York: Farrar, Straus and Giroux, 2022). These notions stem from Christianity and also Enlightenment-era philosophy, in which figures like René Descartes divided the mind from the body and implied a subordinate relationship of the sensory body to the rational brain. In more recent times, this view of the nervous system grew from colonial conquest in the nineteenth century.

28 *territorial expansion overseas:* Vejas Gabriel Liulevicius, "History of the Railroad and Telegraph Binding the World," Wondrium Daily, September 15, 2020, https://www.wondriumdaily.com/history-of-the-railroad-and-telegraph-binding-the-world/.

28 *the main "brain":* John Papiewski, The Similarities between the Human Brain and a CPU," Techwalla, https://www.techwalla.com/articles/the-similarities-between-the-human-brain-and-a-cpu.

33 *"the mysterious butterflies of the soul":* Ehrlich, *The Brain in Search of Itself*, 4.

33 *crucial to our survival:* Ehrlich, *The Brain in Search of Itself*, 186.

33 *Diffusion tensor imaging, or DTI:* For more on diffusion tensor imaging, see Miriam Bolen-Fitzgerald, *Pictures of the Mind: What the New Neuroscience Tells Us About Who We Are* (Upper Saddle River, NJ: Pearson Education, 2010).

33 *significant differences in the tract connections:* For more on differing white matter connectivities in PTSD anxiety and central sensitivity syndrome, see Negar Fani et al., "White matter integrity in highly traumatized adults with and without post-traumatic stress disorder," *Neuropsychopharmacology* 37, no. 12 (2012), 2740–6, doi:10.1038/npp.2012.146.

34 *The Body Keeps the Score:* Bessel van der Kolk, *The Body Keeps the Score* (New York: Viking, 2014).

34 *Matthew Cobb:* Pinkerton, "How technology has inspired neuroscientists to reimagine the brain."

35 *The Brain That Changes Itself:* Norman Doidge, *The Brain That Changes Itself: Stories of Personal Triumph from the Frontiers of Brain Science* (New York: Penguin Life, 2007).

Body Language

37 *home island of Puerto Rico:* Unlike the Philippines, the US has not granted Puerto Rico its independence. Because of this and a history of Spanish colonization that dates back to 1508, Puerto Rico is the oldest colony in the world. For more, see César J. Ayala and Rafael Bernabe, *Puerto Rico in the American Century: A History Since 1898* (Chapel Hill: The University of North Carolina Press, 2007).

37 *Greater United States:* The "Greater United States" is a term coined by historian Daniel Immerwahr. It is a linguistic frame that exposes what the United States has long tried to hide: American imperialism and its impacts on the land and people of its colonies, some of which have since become states: the Philippines, Puerto Rico, Guam, American Samoa, the US Virgin Islands, the Pacific outlying islands, the Caribbean outlying islands, Alaska, and Hawaii. Daniel Immerwahr, *How to Hide an Empire: A History of the Greater United States* (New York: Farrar, Straus and Giroux, 2019), 9.

41 *oranges from Spain:* Besanya Santiago, "The Words and History That Make Puerto Rican Spanish Unique," JP Linguistics, August 13, 2020, https://www.jplinguistics.com/spanish-blog/the-words-and-history-that-make-puerto-rican-spanish-unique.

42 *"embodied flashbacks":* One of the hallmark symptoms of PTSD is the experience of flashbacks, which are defined as a reliving of the traumatic event as if it were happening again. Flashbacks are one subset of a larger category of intrusive memories, which include distressing memories, dreams, and nightmares. See "Post-traumatic stress disorder (PTSD)," Mayo Clinic, https://www.mayoclinic.org/diseases-conditions/post-traumatic-stress-disorder/symptoms-causes/syc-20355967.

42 *Implicit memory:* Patricia Bauer and Jessica A. Dugan, "Chapter 18: Memory development" in *Neural Circuit and Cognitive Development*, 2nd ed., ed. John Rubenstein et al. (Cambridege, MA: Academic Press, 2020), 395–412.

42 *It is habitual:* Paul Renn, "Memory, Trauma and Representation in Psychotherapy," Therapy Route, https://www.therapyroute.com/article/memory-trauma-and-representation-in-psychotherapy-by-p-renn.

42 *reenactment of a preverbal memory:* In an author interview I conducted with Pat Ogden from March 25, 2022, she explained, "The body really holds with its movement and its sensation that which we don't consciously remember."

42 *the age of seven:* Rebecca L. Gómez and Jamie O. Edgin, "The extended trajectory of hippocampal development: Implications for early memory development and disorder," *Developmental Cognitive Neuroscience* 18 (2016), 57–69, doi: 10.1016/j.dcn.2015.08.009.

42 *long-term memories out of storage:* "Brain's memory center stays active during 'infantile amnesia,'" *ScienceDaily*, May 21, 2021, https://www.sciencedaily.com/releases/2021/05/210521115342.htm.

42 *modulate our amygdala's response:* Joseph LeDoux, "The Emotional Brain, Fear, and the Amygdala," *Cellular and Molecular Neurobiology* 23 (2003), 212, doi: 10.1023/A:1025048802629.

43 *because it helps us survive:* Louis Cozolino, *The Neuroscience of Psychotherapy: Building and Rebuilding the Human Brain* (New York: W. W. Norton & Company, 2002), 82–86.

43 *"retention without remembering":* Lisa Firestone, "These Invisible Memories Shape Our Lives," PsychAlive, https://www.psychalive.org/making-sense-of-implicit-memories/.

43 *when they are home alone:* Joshua Kendall, "Forgotten Memories of Traumatic Events Get Some Backing from Brain-Imaging Studies," *Scientific American*, April 6, 2021, https://www.scientificamerican.com/article/forgotten-memories-of-traumatic-events-get-some-backing-from-brain-imaging-studies/.

44 *Jean Clough:* Jean Clough is the cofounder and owner of Seattle Advanced Bodywork Associates (SABA). For more, see "Practitioners at SABA," Seattle Advanced Bodywork Associates (SABA), https://www.seattleadvancedbodywork.com/practitioners/28-jean-clough-lmp.

44 *Bonewhisperer:* The Bone Whisperer is the DBA name of Brian Dobbs, who has developed a unique approach to healing working with what he calls the "fluid tensegrity of the body." As of this printing, Brian Dobbs is working on a book about his method, which has been the most important form of manual therapy in my healing journey. For more, see https://bonewhisperer.net/.

46 *Fugees song:* Fugees, "Killing Me Softly With His Song," track 8 on *The Score*, Columbia Records, 1996.

48 *Black and Native women:* Hill, Latoya, et al. "Racial Disparities in Maternal and Infant Health: Current Status and Efforts to Address Them." *KFF*, 1 Nov. 2022, https://www.kff.org/racial-equity-and-health-policy/issue-brief/racial-disparities-in-maternal-and-infant-health-current-status-and-efforts-to-address-them/.

48 *LGBTQIA+:* Stephanie A. Leonard et al., "Sexual and/or gender minority disparities in obstetrical and birth outcomes," *American Journal of Obstetrics and Gynecology* 226, no. 6 (2022). 846.e1–846.e14. doi:10.1016/j.ajog.2022.02.041.

48 *physical injury impacts:* Apeksha Chaturvedi et al., "Mechanical birth-related trauma to the neonate: An imaging perspective," *Insights into Imaging* 9, no.1 (2018), 103–18, doi:10.1007/s13244-017-0586-x.

49 *African American infants:* A. Jain et al., "Injury related infant death: the impact of race and birth weight," *Injury Prevention* 7, no. 2 (2001), 135–40. doi:10.1136/ip.7.2.135.

Omission

51 *I had a childhood:* This is a nod to Bassey Ikpi's memoir, *I'm Telling the Truth, but I'm Lying.*

53 *my ridiculous needs:* On the psycho-biological mechanism of neglected children who learn to stop reaching out: "We evolved as a species to connect and coregulate through vocal cues, facial gestures, etc. When these cues get disrupted, the natural response is avoidance or shut down." Stephen Porges, in discussion with the author, July 1, 2019.

54 *"narrative elaboration":* Allyssa McCabe, Carole Peterson, and Dianne M. Connors, "Attachment security and narrative elaboration," *International*

Journal of Behavioral Development 30, no. 5 (2006) pp. 398–409. doi:10.1177/0165025406071488.

56 *peddled poisoned cookies:* Granted, my mother's fear of poisoned cookies was not just in her head; blame Dear Abby. Poisoned candy and razor blades in apples were a widespread urban legend in the 1980s and '90s. And yes, my mother was a devoted fan of Dear Abby. For more, see Dan Lewis, "Where Did the Fear of Poisoned Halloween Candy Come From?," *Smithsonian Magazine*, October 6, 2013, https://www.smithsonianmag.com/arts-culture/where-did-the-fear-of-poisoned-halloween-candy-come-from-822302/.

56 *expect to die young:* A foreshortened sense of life is a hallmark symptom of PTSD. See Matthew Tull, "How to Cope With a Sense of a Foreshortened Future," Verywell Mind, April 14, 2022, https://www.verywellmind.com/coping-with-a-foreshortened-future-ptsd-2797225.

56 *after the falls:* At five, I snapped my ankle on a tetherball rope that our gym teacher swung in a circle close to the ground, like "helicopter blades," he said, and at thirteen, I fell down concrete stairs and found out much later that I likely broke my tailbone.

57 *Our Bodies, Ourselves:* The edition of *Our Bodies, Ourselves* that my mother gave me was probably the fifth edition published in 1984. Now there is an interactive website, launched in 2022: https://www.ourbodiesourselves.org/.

58 *I kept mostly to myself:* In an author interview I conducted with Pat Ogden from March 25, 2022, she explained that when proximity-seeking actions are not met by caregivers, we stop reaching out. We wisely stop seeking proximity to others when we haven't gotten what we needed.

58 *Bret Lyon:* Bret Lyon is the cofounder of the Center for Healing Shame. He was the first to help me realize what poet Vanessa Villereal has described: "Trauma is commonly defined as 'too much, too fast' but in my case, it was also 'never enough for too long.'"

60 *thought that something was me:* In this moment, I was making a key shift described by many trauma therapists—the shift from asking what was wrong with me to understanding what had happened to me.

60 *scientists at McGill University:* Michael J. Meaney and Moshe Szyf, "Environmental programming of stress responses through DNA methylation: life at the interface between a dynamic environment and a fixed genome," *Dialogues in Clinical Neuroscience* 7, no. 2 (2005), 103–23, doi: 10.31887/DCNS.2005.7.2/mmeaney.

60 *changed the biochemistry:* Specifically, they found distinct differences in DNA methylation at the NR3C1 promoter in the hippocampus, which can directly impact stress physiology functioning of the HPA axis. See Ian C. G. Weaver et al., "Epigenetic programming by maternal behavior," *Nature Neuroscience* 7, no. 8 (2004), 847–54, doi:10.1038/nn1276.

60 *modifications became lasting changes:* These experiments conducted by Michael Meaney, Moshe Szyf, and Ian Weaver were some of the first to show that genetic expression could be altered by environmental experience. See Bob

Weinhold, "Epigenetics: The Science of Change," *Environmental Health Perspectives* 114, no. 3 (2006), A160–67, doi: 10.1289/ehp.114-a160.

60 *interaction with a caregiver:* Responsive interaction with a caregiver is the basis of the psychological bond known as attachment. Secure attachment bonds have been shown to be a buffer against PTSD and C-PTSD later in life. The interactions that lead to secure attachments also help the development of our brains by teaching our autonomic nervous systems how to regulate themselves. See Sarah Benamer and Kate White, *Trauma and Attachment* (London: Karnac Books, 2008), and Cozolino, *The Neuroscience of Psychotherapy*, 120.

60 *vocalization:* For more on the importance of vocalization and vocal prosody in regulating the nervous system, see Porges, *The Polyvagal Theory*.

60 *"still face":* Mary Gregory, "What does the 'still face' experiment teach us about connection?," PsychHelp, https://www.psychhelp.com.au/what-does-the-still-face-experiment-teach-us-about-connection/.

60 *In a video:* Edward Tronick, "Still Face Experiment Dr Edward Tronick," TNCourts, posted November 9, 2016, YouTube video, https://www.youtube.com/watch?v=IeHcsFqK7So.

61 *"serve and return,":* "Serve and Return," Center on the Developing Child at Harvard University, https://developingchild.harvard.edu/science/key-concepts/serve-and-return/.

61 *Dr. Megan Gunnar:* "Neglect," Center on the Developing Child at Harvard University, https://developingchild.harvard.edu/science/deep-dives/neglect/.

61 *in any tongue:* This is a reference to the fact that I wish my parents had taught me Tagalog, and my mother believes I blame them for neglecting to do so. But I understand this language loss as a broader result of American colonial education, and I don't blame them for it at all. What I do wish is that we spoke more of a language of connection.

62 *help them survive:* "Serve and Return."

62 *illness and substance use:* On the relationship between adverse childhood experiences (ACEs) and long-term health effects, a great source is:

Nadine Burke Harris, *The Deepest Well: Healing the Long-Term Effects of Childhood Trauma and Adversity* (New York: Houghton Mifflin Harcourt, 2018). See also "Long-Term Consequences of Child Abuse and Neglect," Child Welfare Information Gateway, April 2019, p. 9, https://www.childwelfare.gov/pubpdfs/long_term_consequences.pdf.

62 *a predisposing factor [for PTSD]:* Christin M. Ogle, David C. Rubin, and Ilene C. Siegler, "The relation between insecure attachment and posttraumatic stress: Early life versus adulthood traumas," *Psychological Trauma: Theory, Research, Practice and Policy* 7, no. 4 (2015), 324–32, doi:10.1037/tra0000015.

62 *and complex PTSD:* Marloes B. Eidhof et al., "Complex Posttraumatic Stress Disorder in Patients Exposed to Emotional Neglect and Traumatic Events: Latent Class Analysis," *Journal of Traumatic Stress* 32, no. 1 (2019), 23–31, doi:10.1002/jts.22363.

62 *"the silent abuser":* Enod Gray, *Neglect—The Silent Abuser: How to Recognize and Heal from Childhood Neglect* (Houston: TrueSelf Transitions, 2019).

63 *neuromuscular developmental milestones:* For more on crawling and spinal curves, see Stefano Sinicropi, "Infant Spine Development – From the "C" Curve to the "S" Curve," April 25, 2016, https://sinicropispine.com/infant-spine-development-c-curve-s-curve/.

I learned of the connection between failure to crawl, scoliosis, and pain thanks to one of my first chiropractors, Dr. Amelia Mazgaloff. Dr. Mazgaloff is the founder of Chiro-Health in downtown San Francisco. See "Dr. Amelia Mazgaloff," Chiro-Health Chiropractic, https://www.chirohealthsf.com/our-staff/dr-amelia-mazgaloff/.

I learned about neuromuscular developmental milestones and their importance to motor control and pain prevention from one of my most recent chiropractors, Dr. Krysann Rodriguez at Tangelo Health in Seattle. See https://www.tangelohealth.com/seattle-chiropractor/. Dr. Krysann applied a technique called Dynamic Neuromuscular Stabilization to help reestablish this neuromuscular stabilization I had otherwise missed out on because of neglect. See https://www.rehabps.com/.

63 *neglect is a deeper level:* Gray defines neglect as a persistent pattern in which a child's physical and/or emotional needs are ignored by caregivers. Gray, *Neglect—The Silent Abuser*.

63 *easy to dismiss:* According to human development expert Brenda Jones Harden, children are much more likely to experience neglect than they are to experience any other kind of abuse. For more, see "Neglect," *Center on the Developing Child at Harvard University*. Of children who are subjected to maltreatment one out of ten experience sexual abuse, one out of ten face physical abuse, and six out of ten are neglected. Because children who experience physical and/or sexual abuse from primary caregivers are also bound to be neglected, neglect is considered a cornerstone form of mistreatment that precedes all other forms of abuse. For more, see "Child abuse, neglect data released," Administration for Children & Families, January, 15, 2020, https://www.acf.hhs.gov/media/press/2020/2020/child-abuse-neglect-data-released.

63 *emotional needs are on par:* It turns out that emotional connection with a caregiver may be more important than material sustenance, at least for monkeys. In a famous experiment known as Harlow's monkey experiment, behavioral scientist Harry Harlow constructed two fake monkey mothers out of wire. They draped one of the wire mothers with terry cloth, and the other they left bare but equipped it with milk. Which of the surrogate monkeys did the infants choose? "The results were unambiguous and profound," reports neuroscientist Matthew Lieberman. The infant monkeys spent nearly eighteen hours a day in contact with the terry cloth–draped monkey. "These monkeys were attached to the thing that felt most like a real monkey, regardless of the sustenance it provided." See Matthew D. Lieberman, *Social: Why Our Brains Are Wired to Connect* (New York: Crown Publishers, 2013), 49. See also

"Harlow's Classic Studies Revealed the Importance of Maternal Contact," Association for Psychological Science—APS, June 20, 2018, https://www.psychologicalscience.org/publications/observer/obsonline/harlows-classic-studies-revealed-the-importance-of-maternal-contact.html.

Mobility

66 *University of the Philippines:* The University of the Philippines is a public university that was started by the American colonial government. The Diliman campus in Quezon City is the largest, but UP has campuses all over the islands. For more, see "University of the Philippines," PhilippineUniversity Wiki, https://phcollegesuniv.fandom.com/wiki/University_of_the_Philippines. UP was one example of how the United States molded the Philippines in its image through education and culture but also through the built environment. For more on this, see Rebeca Tinio McKenna, *American Imperial Pastoral: The Architecture of US Colonialism in the Philippines* (Chicago: The University of Chicago Press, 2017).

67 *Quonset hut:* Edgar Allan M. Sembrano,"Wartime quonset hut in UP Diliman demolished," *Philippine Daily Inquirer*, December 12, 2016, https://lifestyle.inquirer.net/247272/wartime-quonset-hut-diliman-demolished/.

67 *offering themselves up to the nation:* Oliver Carlos, "Ten Facts About the UP Oblation," *Medium*, December 12, 2021, https://oliverjetcastillo.medium.com/ten-facts-about-the-up-oblation-1582aa05fd5d.

67 *facing west:* "The Oblation Unveiled in 1935," Bahay Nakpil-Bautista, https://bahaynakpil.org/the-oblation-unveiled-in-1935/.

67 *"Greater United States":* Immerwahr, *How to Hide an Empire.*

68 *Tondo, Manila:* For a legendary fictional account of Tondo, written by an author who went to the same high school as my father, see Andres Cristobal Cruz, *Ang Tundo Man May Langit Din* (Manila: Ateneo de Manila University Press, 1986).

68 *breeding revolutionaries:* Wilson Lee Flores, "Proud to be a Tondo boy," Philstar.com, February 20, 2005, https://www.philstar.com/lifestyle/Sunday-life/2005/02/20/268792/proud-be-tondo-boy.

68 *Andrés Bonifacio:* Manuel Pardo, "Know Your Heroes: Andres Bonifacio," Filipino Association of Greater Kansas City, June 2016, http://www.filipino-association.org/tambuli/showarticle.asp?_id=90.

68 *Marina Dizon:* "Marina Dizon," Wikipedia, last modified December 4, 2021, https://en.wikipedia.org/w/index.php?title=Marina_Dizon&oldid=1058552565.

68 *Macario Sakay:* "Macario Sakay," Bayani Art, https://www.bayaniart.com/articles/macario-sakay-biography/.

68 *people moved from rural areas:* Carlos Celdran, "Tondo: Manila's Largest Slum - The Space in Between," Al Jazeera English, posted September 10, 2014. YouTube video, https://www.youtube.com/watch?v=OIGA0xDDagQ.

68 *violence from street gangs:* Karl Aguilar, "At the heart of Tondo," *The Urban Roamer* (blog), May 10, 2018, https://www.theurbanroamer.com/at-the-heart-of-tondo/.

68 *sipa, patintero, and trumpo:* Kim Shelly Tan, "11 Filipino Childhood Games That Made Our Summers Fun, From Teks to Patintero," *The Smart Local Philippines*, March 11, 2022, https://thesmartlocal.com/philippines/filipino-childhood-games/.

68 *Plaza Moriones:* Plaza Moriones is one of the most historically significant places in Tondo. For more on its history, see Karl Aguilar, "A Tale of Two Tondo Plazas," *The Urban Roamer* (blog), May 18, 2018, https://www.theurbanroamer.com/tale-of-two-tondo-plazas/. For a picture of Plaza Moriones in the 1950s, see Nitoy Ibanaz, "Pinoy," Pinterest, https://www.pinterest.com/pin/303078249896703520/.

68 *Plaza Miranda:* "Plaza Miranda," Wikipedia, last modified July 14, 2022, https://en.wikipedia.org/w/index.php?title=Plaza_Miranda&oldid=1098150929.

68 *Divisoria:* "Divisoria," Wikipedia, last modified September 1, 2022, https://en.wikipedia.org/w/index.php?title=Divisoria&oldid=1107858720.

68 *sabong:* Ronald de Jong, "Sabong, 'The Sport of Kings,'" *ThingsAsian*, March 1, 2013, http://thingsasian.com/story/sabong-%E2%80%9C-sport-kings%E2%80%9D.

69 *flag down a jeepney:* For a brief history of the jeepney, see Roy Robles, "Arts & Culture: A Look into the History of Jeepneys — the Fragile Kings of the Philippine Roads," *Adobo Magazine*, November 28, 2019, https://www.adobomagazine.com/philippine-news/arts-culture-a-look-into-the-history-of-jeepneys-the-fragile-kings-of-the-philippine-roads/.

69 *upward mobility for the Filipino masses:* Gregorio Lim (Family Planning Association of the Philippines), in discussion with the author, May 15 , 1997.

In this interview, Lim provided me with a detailed story of the origins of the family planning movement in the Philippines, details on the process of contraceptive importation, and on deteriorating conditions in Tondo as a motivating force for family planning initiation.

69 *access to birth control:* Ana Maria Nemenzo (WomanHealth Philippines), in discussion with the author, August 11, 1997.

In this interview, Nemenzo traced the history of Filipina women's activism surrounding contraceptives.

69 *and still do:* Susan M. Blaustein, "Progress Towards Contraceptive Access in the Philippines," *Ms.*, July 20, 2020,. https://msmagazine.com/2020/07/20/progress-towards-contraceptive-access-in-the-philippines/.

69 *coercive population control:* Donald Warwick, *Bitter Pills: Population Policies and their Implementation in Eight Developing Countries* (Cambridge, MA: Cambridge University Press, 1982).

70 *colonial mentality:* Colonial mentality in Filipino American communities has been defined and extensively studied by first-generation Filipino American

psychologist E. J. R. David. For more, see his seminal book on the subject: E. J. R. David, *Brown Skin, White Minds: Filipino-American Postcolonial Psychology* (Charlotte, NC: Information Age Publishing, 2013).

71 *great circle:* "Great-circle distance," Wikipedia, last modified 5 September 5, 2022, https://en.wikipedia.org/w/index.php?title=Great-circle_distance&oldid=1108637073.

73 *twenty million dollars:* Immerwahr, *How to Hide an Empire*.

73 *"The White Man's Burden":* For more on Kipling, see Immerwahr, *How to Hide an Empire*, 94. On McKinley, see Immerwahr, *How to Hide an Empire*, 74.

73 *"little brown brothers":* William Howard Taft, who was appointed as the Philippines's first governor-general under the American colonial government, coined this term when reporting to President McKinley on the amount of time it would take to civilize the Filipino people. For more, see "Little brown bother," Wikipedia, last modified May 23, 2022, https://en.wikipedia.org/wiki/Little_brown_brother. For a pictorial history of America's view of "little brown brothers" in political cartoons, see Abe Ignacio et al., *The Forbidden Book: The Philippine-American War in Political Cartoons* (San Francisco: T'Boli Publishing and Distribution, 2004).

73 *William Morgan Shuster:* "University History," University of the Philippines, https://up.edu.ph/university-history/.

73 *state universities of California:* University of the Philippines, *A brief history of the University of the Philippines*, ed. Felipe Estella (Manila: University of the Philippines, 1922), http://name.umdl.umich.edu/AGE4844.0001.001.

74 *family reunification:* Barbara M. Posadas, *The Filipino Americans* (Westport, CT: Greenwood Press, 1999).

74 *industry in America was growing:* "A History of the Pharmaceutical Industry," *pharmaphorum*, 1 September 1, 2020, https://pharmaphorum.com/r--d/a_history_of_the_pharmaceutical_industry/.

74 *grew significantly after 1965:* Catherine Ceniza Choy, *Empire of Care: Nursing and Migration in Filipino American History* (Durham, NC: Duke University Press, 2003).

75 *Exposition:* The World's Fair and the Space Needle were demonstrations of Seattle's growth as an aerospace city with the power to help the United States win the Cold War. For more, see "Century 21 World's Fair," Seattle Municipal Archives, https://www.seattle.gov/cityarchives/exhibits-and-education/digital-document-libraries/century-21-worlds-fair.

78 *spinal cord problems:* Danilo Soriano et al., "Experimental Relief of Spasticity in the Cat by Lumbosacral Radio-Frequency Cordotomy," *Transactions of the American Neurological Association* 93 (1968).

78 *requirement of his J-1 visa:* "Immigration: Foreign Physicians and the J-1 Visa Waiver Program," EveryCRSReport.com, December 9, 2004, https://www.everycrsrepor.com/reports/RL31460.html.

79 *There were no grandparents:* Anthony Ocampo writes about how Filipinos often don't have to live in ethnic enclaves because of their acculturation to

American English and culture. See Anthony Christian Ocampo, *The Latinos of Asia: How Filipino Americans Break the Rules of Race* (Stanford, CA: Stanford University Press, 2016), 31. My parents were able to be so much "of this place" that they were able to go it almost completely alone. Many Filipinos in the US live with what Rhacel Salazar Parreña calls an extended rather than nuclear family base. But our family was an exception. See Rhacel Parreñas, *Children of Global Migration: Transnational Families and Gendered Woes* (Stanford, CA: Stanford University Press, 2005).

79 *model of white suburban America:* "Dinner with the nuclear family, 1950," Gilder Lehrman Institute of American History, https://www.gilderlehrman.org/history-resources/spotlight-primary-source/dinner-nuclear-family-1950.

80 *she helped:* For more on domestic workers "being in the middle" of middle-class employers and those too poor to migrate, see Rhacel Salazar Parreñas, "Migrant Filipina Domestic Workers and the International Division of Reproductive Labor," *Gender and Society* 14, no. 4 (2000), 560–80. For more on the intertwining of employers and workers' American Dreams, see Susanna Rosenbaum, *Domestic Economies: Women, Work, and the American Dream in Los Angeles* (Durham, NC: Duke University Press, 2017).

80 *sent much of her wages:* In 2021, Filipino overseas workers sent more than thirty-one billion dollars in remittances to family in the Philippines. See Ismaeel Naar, "Cash remittances from overseas workers remain a lifeline for many in the Philippines," *The National*, June 16, 2022, https://www.thenationalnews.com/world/asia/2022/06/16/cash-remittances-from-overseas-workers-remain-a-lifeline-for-many-in-the-philippines/. These remittances make up a significant portion of the Philippine GDP. See "Philippines - Remittance Inflows To GDP," Trading Economics, https://tradingeconomic.com/philippines/remittance-inflows-to-gdp-percent-wb-data.html.

80 *four million:* The estimate of four million Filipina domestic workers comes from the numbers in this article:

Corinne Redfern, "'I want to go home': Filipina domestic workers face exploitative conditions," *The Guardian*, January 27, 2021, https://www.theguardian.com/world/2021/jan/27/domestic-workers-philippines-coronavirus-conditions.

80 *Lydia sacrificed:* For more on Filipino domestic workers, their sacrifices, struggles, and vital role in the global economy, see Rhacel Salazar Parreñas, *Servants of Globalization: Migration and Domestic Work* (Stanford, CA: Stanford University Press, 2001).

80 *Perhaps colonial mentality:* For more on the effects of colonial mentality on Filipino American immigrants, as well as a critique of the voluntary immigrant narrative when coming from an American colony, see E. J. R. David and Kevin L. Nadal. "The Colonial Context of Filipino American Immigrants' Psychological Experiences," *Cultural Diversity and Ethnic Minority Psychology* 19, no. 3 (2013), 298–309, doi:1037/a0032903.

81 *patók jeepney:* Fame Pascua, "The *Patók* Jeep: A Descriptive Study," *Asian*

Studies 45, nos. 1–2 (2009), 83–92, https://www.asj.upd.edu.ph/mediabox/archive/ASJ-45-1and2-2009/pascua.pdf

82 *Separation from others:* Colonial mentality may have played a role in this separation. "The belief or perception that their group is inferior may lead individuals to isolation if they feel inclined to separate themselves from their family or community because they're ashamed of them, he said. And that can leave people without a buffer—something that's critical during challenging times." E. J. R. David, as told to Agnes Constante, "How the Philippines' Colonial Legacy Weighs on Filipino American Mental Health," *Los Angeles Times*, October 12, 2021, https://www.latimes.com/lifestyle/story/2021-10-12/colonial-history-behind-filipino-american-mental-health.

82 *balikbayan boxes:* daleasis, "Balikbayan Boxes: Symbols of Homesickness, Colonial History, and Family," Bayanihan Foundation Worldwide, October 27, 2018, https://fdnbayanihan.org/2018/10/27/balikbayan-boxes-symbols-of-homesickness-colonial-history-and-family/.

War-Fire

88 *just ten hours:* John T. Correll, "Disaster in the Philippines," *Air & Space Forces Magazine*, November 1, 2019, https://www.airandspaceforces.com/article/disaster-in-the-philippines/.

88 *caught unprepared:* Immerwahr, *How to Hide an Empire*, 4.

93 *Maria Yellow Horse Brave Heart:* Professor Maria Yellow Horse Brave Heart developed the concept of historical trauma, as well as a framework and approach for intervention among Native communities. This is one of many papers she was written on the topic: Maria Yellow Horse Brave Heart, "*Wakiksuyapi*: Carrying the historical trauma of the Lakota," *Tulane Studies in Social Welfare* 21–22 (2000), 245–66.

93 *the roots of trauma:* For a model of how historical trauma may be biologically transmitted through epigenetic modifications (as opposed to just through behavior and narratives), see Andie Kealohi Sato Conching and Zaneta Thayer, "Biological pathways for historical trauma to affect health: A conceptual model focusing on epigenetic modifications," *Social Science & Medicine* 230 (2019), 74–82, doi:10.1016/j.socscimed.2019.04.001.

93 *unresolved generational grief:* "Maria Yellow Horse Brave Heart: Historical Trauma and Healing in Native Communities," Healing Collective Trauma, https://www.healingcollectivetrauma.com/dr-maria-yellow-horse-brave-heart-historical-trauma-in-native-communities.html.

93 *a transgenerational perspective:* For Maria Yellow Horse Brave Heart's teaching on intergenerational healing, see "From Intergenerational Trauma to Intergenerational Healing," *Wellbriety! White Bison's Online Magazine* 6, no. 6, (2005), https://www.sjsu.edu/people/marcos.pizarro/maestros/BraveHeart.pdf.

94 *trauma in her environment:* "*Epigenetics: The Hidden Life of Our Genes.*" Filmed

2012. Infobase, 8. https://seattle.bibliocommons.com/v2/record/S30C3017642.

94 *sensitive time for neural pruning:* "Sensitive Periods," Better Brains for Babies, https://www.bbbgeorgia.org/sensitive-periods.

95 *mother's eggs have been altered:* Young people who have survived famine go on to have children with metabolic disorders like hypoglycemia, perhaps because of changes to the parents' egg and sperm triggered by nutritional deprivation. These changes may be specifically epigenetic, which means they are changes to the way a gene is expressed rather than a change to the genes themselves. They occur as adaptations to new environments, like a sudden and prolonged lack of food, or the coming of war, and can affect how certain genes are expressed in the next generations. While many of these changes are "scrubbed clean" when an embryo is formed, some may be so important to survival that they remain and are passed on.

Vicencia Micheline Sales, Anne C. Ferguson-Smith, and Mary-Elizabeth Patti, "Epigenetic Mechanisms of Transmission of Metabolic Disease across Generations," *Cell Metabolism* 25, no. 3 (2017), 559–71, doi:10.1016/j.cmet.2017.02.016.

K. Gapp and J. Bohacek, "Epigenetic germline inheritance in mammals: looking to the past to understand the future," *Genes, Brain and Behavior* 17, no. 3 (2018), e12407, doi:10.1111/gbb.12407.

Bianca Jones Marlin (neuroscientist, Columbia University), in discussion with the author, February 2, 2022.

97 *More than 1.1 million:* Jeffrey Hays, "Defeat of Japan in the Philippines," *Facts and Details (blog),* last modified November 2016, https://factsanddetails.com/asian/ca67/sub428/item2522.html.

97 *eight guerrilla units:* James Kelly Morningstar, *War and Resistance in the Philippines 1942–1944* (Annapolis, MD: Naval Institute Press, 2021).

98 "*trauma ghosting*": Resmaa Menakem, *My Grandmother's Hands: Racialized Trauma and the Pathway to Mending Our Hearts and Bodies* (Las Vegas: Central Recovery Press, 2017), 8.

99 *memorial candles:* Dina Wardi, *Memorial Candles: Children of the Holocaust* (New York: Routledge, 1992).

99 *survivors themselves:* "Rachel Yehuda — How Trauma and Resilience Cross Generations," July 30, 2015, in *On Being with Krista Tippett*, podcast, MP3 audio, https://soundcloud.com/onbeing/rachel-yehuda-how-trauma-and-resilience-cross-generations-nov2017.

99 *more likely to have PTSD:* Rachel Yehuda et al., "Holocaust Exposure Induced Intergenerational Effects on FKBP5 Methylation," *Biological Psychiatry* 80, no. 5 (2016), 372–80, doi:10.1016/j.biopsych.2015.08.005.

99 *Holocaust as children:* Linda M. Bierer et al., "Intergenerational Effects of Maternal Holocaust Exposure on *FKBP5* Methylation," *The American Journal of Psychiatry* 177, no. 8 (2020), 744–53, doi:10.1176/appi.ajp.2019.19060618.

100 *traumatic retention:* Resmaa Menakem's work was key to my understanding of the difference between culture and traumatic retention. Mankem, *My Grandmother's Hands*.

103 *four generations later:* For Coleen Murphy's team's original journal article on epigenetic transmission of avoidance across four generations of *C. elegans*, see Rebecca S. Moore, Rachel Kaletsky, and Coleen T. Murphy, "Piwi/PRG-1 Argonaute and TGF-β Mediate Transgenerational Learned Pathogenic Avoidance," *Cell* 177, no. 7 (2019), 1827–1841.e12, doi:10.1016/j.cell.2019.05.024; and a subsequent publication, Rebecca S. Moore et al., "Horizontal and vertical transmission of transgenerational memories via the *Cer1* transposon," bioRxiv, December 29, 2020, doi:10.1101/2020.12.28.424563.

For a journalistic narrative of the same, see Caitlin Sedwick, "Murphy Lab researchers discover how worms pass knowledge of a pathogen to offspring," Princeton University Department of Molecular Biology, September 9, 2020, https://molbio.princeton.edu/news/murphy-lab-researchers-discover-how-worms-pass-knowledge-pathogen-offspring.

104 *This sense of neuroception:* Stephen W. Porges, "Polyvagal Theory: A biobehavioral journey to sociality," *Comprehensive Psychoneuroendocrinology* 7 (2021), 100069, doi:10.1016/j.cpnec.2021.100069.

105 *fifth generation:* Coleen Murphy, in discussion with the author, December 9, 2020.

105 *can eventually be reversed:* "You can write a CD but you can't erase it, but then you run out of space. But DNA methylation is like a flash drive, you can write and rewrite—at the biological level, it is these type of epigenetic changes that allow encoding of plasticity." Hongjun Song (director of neuroepigenetics, University of Pennsylvania), in discussion with the author, November 13, 2020.

105 *For humans:* On extrapolation of worm studies to humans:"It's natural to extrapolate. It would be surprising if worms had mechanisms for transgenerational inheritance that no other animal does. What's to stop another animal with the same kind of genes from doing a similar thing?" Murphy, discussion.

381 Years

106 *381 Years:* 381 years is a calculation of the total years of Spanish and American colonization in the Philippines. The span of time varies depending on how you define when colonization began. Colloquial estimates say 350 years, which probably comes from the saying that Filipinos spent 300 years in a convent and 50 years in Hollywood. Some count from 1521 when Ferdinand Magellan first landed on Cebu. This would lead to a total of 377 years of Spanish colonization plus 47 years of American rule for a total of 424 years. I count from 1565, the year that Miguel Lopez de Legazpi established the first Spanish settlement in Cebu. This acknowledges the successful resistance against Spaniards' arrival in 1521, including Lapu Lapu's felling of

Magellan, as well as the fact that it took conquistadors another few decades to successfully return to the islands. From 1565 to 1946, when the United States officially granted independence to the Philippines, equals 381 years of colonization.

A note on the form: recounting 381 years of colonization, resistance to colonization, and the embodiment of its impacts seemed impossible to do in one essay in traditional narrative form. I needed a form that could present multiple layers and perspectives. I thought of hip-hop and its tradition of call and response, which also made me think of the Tagalog tradition of Balagtasan poetry, a form of poetic competition that resembles rap battles. But how could I set that up on the page? I tried out what I thought of as a Bible format, phrases or verses laid out in columns. And then I recalled the metaphor of counterpoint, developed by D.R.M. Irving, in which he breaks down how the Spanish musical approach to counterpoint was an instrument of colonialism, insofar as counterpoint involved a rigid set of rules "wielded by a manipulating power." In contrast, contrapuntal literary analysis as proposed by Edward Said, "addresses the perspectives of both the elite and the subaltern, using binary oppositions to create a framework within which meaning can be constructed." Said quoted in D.R.M. Irving, *Colonial Counterpoint* (Oxford: Oxford University Press, 2010), 7.

107 *conquistador Ferdinand Magellan:* Luis H. Francia, *A History of the Philippines: From Indios Bravos to Filipinos* (New York: Overlook Press, 2010), 54.

107 *three goats, three pigs:* "Ferdinand Magellan's Voyage Round the World, 1519-1522 CE," Fordham University Modern History Sourcebook, https://sourcebooks.fordham.edu/mod/1519magellan.asp. This is a primary source transcribed from the paper-book of a Genoese pilot who was on the voyage with Magellan, translated, transcribed and scanned by J.S. Arkenberg, Department of History, California State Fullerton.

107 *warriors felled the conquistador:* Ricardo Espinoza-González et al., "The possible poisons contained in the arrow that killed Ferdinand Magellan," *Revista Medica De Chile* 149, no. 11 (November 2021): 1636–41. https://doi.org/10.4067/S0034-98872021001101636.

107 *three more tries:* Francia, *A History of the Philippines*, 56. In 1543, an expedition led by Roy López de Villalobos landed in Mindanao and started a colony in Sarangani. The Villalobos expedition christened the archipelago "Filipinas."

108 *great success:* Francia, *A History of the Philippines,* 56–60.

108 *the wife of Rajah Humabon:* Antonio D. Sison, "Manila's Black Nazarene and the Reign of Bathala," *Journal of Global Catholicism* 6, no. 1 (December 2021): 64–80. https://doi.org/10.32436/2475-6423.1105.

108 *Legazpi continued:* Nicholas P. Cushner, "Legazpi 1564-1572," *Philippine Studies* 13, no. 2 (1965): 163–206.

109 *regime of repression:* Renato Constantino, *The Philippines Volume 1: A Past Revisited* (Manila, 1975), 44–65.

109 *"docile" and "simple":* James A. LeRoy, "The Friars in the Philippines," *Political Science Quarterly* 18, no. 4 (1903): 657–80. https://doi.org/10.2307/2140780.

109 *half the population:* Anchi Hoh, "Catholicism in the Philippines during the Spanish Colonial Period 1521-1898," 4 Corners of the World: International Collections and Studies at the Library of Congress, July 10, 2018, https://blogs.loc.gov/international-collections/2018/07/catholicism-in-the-philippines-during-the-spanish-colonial-period-1521-1898/.

110 *the patron saint of Philippine revolution:* "Santa Maria Magdalena de Kawit: Kawit's Loving Paraluman," Santa Maria Magdalena de Kawit, September 22, 2016, https://pintakasi1521.blogspot.com/2016/09/santa-maria-magdalena-de-kawit-kawits.html.

110 *made water god:* Sison, "Manila's Black Nazarene and the Reign of Bathala."

111 *a way station:* Constantino, *A Past Revisited,* 55–65.

111 *bodies of war:* Constantino, *A Past Revisited,* 85–150.

111 *native tongues:* For more on why Spain didn't teach Spanish to Filipinos, see Constantino, *A Past Revisited,* 150–70.

111 *appropriated the king's tongue:* See, for example Eeva Sippola, "Ilokano-Spanish: Borrowing, Code-Switching or a Mixed Language?" *Ilokano-Spanish: Borrowing, Code-Switching or a Mixed Language?* (2021): 253–76. https://doi.org/10.1515/9781501511257-009.

111 *agitate to become a nation:* Constantino, *A Past Revisited,* 133–52.

111 *Cry of Balintawak:* Soledad Borromeo-Buehler, *The Cry of Balintawak* (Manila: Ateneo de Manila University Press, 1998).

112 *"Mabuhay ang Pilipinas! Mabuhay ang Pagsasarili!":* Borromeo-Buehler, *The Cry of Balintawak,* 148.

112 *the Philippines rose up*: See Constantino, *A Past Revisited,* 173–204.

113 *"I should welcome almost any war":* Immerwahr, *How to Hide an Empire,* 64.

113 *Spanish-American-Cuban-Filipino War:* Harry Franqui-Rivera, "The Spanish-American-Cuban-Filipino War of 1898," https://centropr-archive.hunter.cuny.edu/digital-humanities/pr-military/spanish-american-cuban-filipino-war-1898.

113 *the arms never appeared:* Agoncillo, *History of the Filipino People,* 98.

114 *heart attack:* Immerwahr, *How to Hide an Empire,* 67.

114 *"ignorant of the location of the Philippines":* Agoncillo, *History of the Filipino People,* 197.

114 *"willing to surrender to white people but never to n——s":* Immerwahr, *How to Hide an Empire,* 72.

114 *twenty million dollars:* Teodoro A. Agoncillo, *History of the Filipino People* (Quezon City: C&E Publishing, 2012), 202–05, 220.

114 *stage a mock-battle:* Agoncillo, *History of the Filipino People,* 202.

115 *"The band struck up 'The Star-Spangled Banner'":* Immerwahr, *How to Hide an Empire,* 72.

115 *an American soldier shot three Filipino soldiers:* Immerwahr, *How to Hide an Empire,* 90.

115 *American clapback was swift:* Agoncillo, *History of the Filipino people*, 226.

116 *"America's first Vietnam":* Luzviminda Francisco, "The First Vietnam: The Philippine-American War, 1899-1902," in *The Philippines Reader: A History of Colonialism, Neocolonialism, Dictatorship, and Resistance*, ed. Daniel B. Schirmer et al., (Boston: South End Press, 1987).

116 *Rudyard Kipling published his famous poem:* Few know that the full title of this poem is "The White Man's Burden: The United States and the Philippine Islands." Teddy Roosevelt leveraged it for imperialist purposes, but called it "rather poor poetry." "'The White Man's Burden': Kipling's Hymn to U.S. Imperialism." http://historymatters.gmu.edu/d/5478.

116 *"The defenses of the colonized are tuned like anxious antennae":* Fanon, *The Wretched of the Earth*, ix. This quote is something Fanon said to Jean-Paul Sartre, as paraphrased by Homi K. Bhabha in the forward to this edition of the book.

116 *a "quail shoot":* Francisco, "The First Vietnam: The Philippine-American War, 1899-1902."

116 *called Filipinos "n——s," "barbarians," and "savages":* For an excellent source on American racism and the role that anti-Blackness and anti-Native American racialization played in justifying the U.S. occupation of the Philippines, see Abe Ignacio, et al. *The Forbidden Book* (San Francisco: T'Boli Publishing and Distribution, 2004).

117 *"It may be necessary to kill half of the Filipinos":* quoted in Francisco, "The First Vietnam: The Philippine-American War, 1899-1902."

117 *hit on the head 125 times:* these are references to some of the most egregious attacks on Asian American women during the surge of anti-Asian hate that happened during the COVID-19 pandemic. Specifically, these phrases refer to the March 16, 2021 murder of six Asian massage parlor workers in Atlanta, two Chinese women and two Korean women, the March 29, 2021 beating of a sixty-five-year-old Filipino woman on her way to church in New York City, and the March 11, 2022 beating of a sixty-seven-year-old Filipino woman at her apartment building in Yonkers.

117 *"The supremacy of the United States must and will be enforced":* Constantino, *A Past Revisited*, 226.

117 *"Americans used the bodies for breastworks":* Francisco, "The First Vietnam: The Philippine-American War, 1899-1902."

118 *An American soldier wrote home:* Francisco, "The First Vietnam: The Philippine-American War, 1899-1902."

118 *guerrilla warfare:* Constantino, *A Past Revisited*, 256–86.

119 *"The success of this unique system of war":* Arthur MacArthur, father of General Douglas MacArthur, quoted in Constantino, *A Past Revisited*, 234.

119 *there were no more civilians:* Francisco, "The First Vietnam: The Philippine-American War, 1899-1902."

119 *An American congressman said:* Francisco, "The First Vietnam: The Philippine-American War, 1899-1902."

120 *Balangiga massacre:* For a powerful account of the Balangiga massacre conveyed through fiction, see Gina Apostol, *Insurrecto* (New York: Soho Press, 2018).

120 *Bud Dajo "dwarfed them all":* Immerwahr, *How to Hide an Empire*, 105–07.

120 *the United States's longest war:* Immerwahr, 107.

121 *six hundred thousand Filipinos in Luzon alone were killed:* Estimation from General Bell as told to the *New York Times* and mentioned in Francisco, "The First Vietnam: The Philippine-American War, 1899-1902."

121 *Our bodies continue to bear the long burden:* There is a growing body of literature on colonization as a risk factor for long-term disease, which builds on research that links adverse childhood experiences (ACEs) to adult disease. For a thorough discussion of ACEs and disease, see Harris, *The Deepest Well: Healing the Long Term Effects of Childhood Adversity*. For qualitative links between colonization and disease see Karina L. Walters, et al., "Bodies Don't Just Tell Stories, They Tell Histories," *Du Bois Review : Social Science Research on Race* 8, no. (April 2011): 179–89. https://doi.org/10.1017/S1742058X1100018X.

121 *more than one million Filipinos died:* Francisco, "The First Vietnam: The Philippine-American War, 1899-1902."

121 *"the bravest men I have ever seen":* General Lawton quoted in Francisco, "The First Vietnam: The Philippine-American War, 1899-1902."

121 *transgenerational bravery:* Professor Valerie Francisco-Menchavez introduced me to the concept of transgenerational bravery in a talk she gave at the Pinay Power Summit at McGill University in April 2019. For more on this concept contact her for the paper "Transgenerational Bravery: Teaching and Taking a Stand Against Empire."

122 *American imperialism was an expansion:* Nerissa S. Balce, *Body Parts of Empire* (University of Michigan Press, 2019), 12.

122 *wield laughter like a weapon:* On the subject of laughter as resistance, see Majken Jul Sorensen, "Humor as a Serious Strategy of Nonviolent Resistance to Oppression," *Peace & Change* 33, no. 2 (2008): 167–90. https://doi.org/10.1111/j.1468-0130.2008.00488.x.

122 *redistributing land held by Catholic friars:* Agoncillo, *History of the Filipino People*, 385.

123 *should not be mistaken for benevolence:* On the complications of American empire and benevolence in the Philippines, see Thomas McCormick, "From Old Empire to New: The Changing Dynamics and Tactics of American Empire," in *Colonial Crucible: Empire in the Making of the Modern American State*, eds. Alfred W. McCoy and Francisco A. Scarano. (Madison: University of Wisconsin Press, 2009).

123 *"little brown brothers":* a paternalistic term used by American Governor General of the Philippines William Howard Taft. https://en.wikipedia.org/wiki/Little_brown_brother.

123 *Filipinos are like chocolate-covered pretzels:* Jason W. Lloren, "Filipino American

Comic Rex Navarrete Combines Insight, Biting Wit," *SFGate*, June 23, 2002. https://www.sfgate.com/bayarea/article/PROFILE-Rex-Navarrete-Filipino-American-2827266.php.

123 "Asia for Asians": Renato Constantino and Letizia Constantino, *The Philippines: The Continuing Past* (Quezon City: The Foundation for Nationalist Studies, 1978).

124 *to avoid fighting Japan:* Agoncillo, *History of the Filipino People*, 402.

124 *conquered and abandoned:* Immerwahr, *How to Hide an Empire*, 189.

124 *"We're the battling bastards of Bataan":* Soldiers sang this song based on a poem by Frank Hewlett. The song may have been slightly different than the poem, but the original poem is quoted here (with thanks to Kaitlyn San Miguel). American Defender of Bataan and Corregidor Museum, https://philippine defenders.pastperfectonline.com/archive/7EE89BDF-A548-4653-A6E3 -388988537800

125 *"the terrific lashings they have received":* Eliseo Quirine, *A Day to Remember*, Manila, 1958, 79, quoted in Immerwahr, 197.

125 *"No one, during the darkest days of the occupation, could sleep soundly":* Agoncillo, *History of the Filipino People*, 420.

125 *internet meme on a Brief History of the Philippines:* posted by taongkalye on *reddit* in the "Polandball"community, 10 Dec. 2013, www.reddit.com/r/po landball/comments/1sjqwn/a_brief_history_of_the_philippines/.

125 *Guerrilla units of resistance once again:* Agoncillo, *History of the Filipino People*, 429.

126 *take back Taiwan:* Immerwahr, 201.

126 *Freud had to write a book:* Sigmund Freud, *Jokes and Their Relation to the Unconscious* (New York: W.W. Norton & Company, 1990).

127 *Japanese Army's escape route:* Some military historians have called this "*the* strategic blunder of the Philippine campaign." Immerwahr, 205.

128 *heavy-handed with its artillery:* Immerwahr, 206.

128 *Americans bombed the Philippine General Hospital:* As well as other refuge centers, Immerwahr, 208.

128 *"there was nothing left":* Immerwahr, 211.

129 *one hundred thousand Manila residents:* "For every 'American life' lost, 100 Manilans died." Immerwahr, 211.

129 *more than 1.6 million*: Francisco, "The First Vietnam: The Philippine-American War, 1899-1902."

129 *"most destructive event ever to take place on US soil":* Immerwahr, 212.

129 *A photo of Manila:* "Battle of Manila (1945)," Wikipedia. https://en.wikipedia .org/w/index.php?title=Battle_of_Manila_(1945)&oldid=1103401446. And "The Americans Destroyed Manila in 1945." *RAPPLER*, 4 Feb. 2015, https:// www.rappler.com/newsbreak/iq/82850-americans-destroyed-manila-1945/.

130 *second only to Warsaw:* Ishaan Thoroor, "Manila was known as 'The Pearl of the Orient.' Then World War II Happened," *The Washington Post*, February 19, 2015. https://www.washingtonpost.com/news/worldviews/wp/2015/02/19 /manila-was-known-as-the-pearl-of-the-orient-then-world-war-ii-happened/.

130 *Warsaw was painstakingly rebuilt:* Daryl Mersom, "Story of Cities #28: How Postwar Warsaw Was Rebuilt Using 18th Century Paintings," *The Guardian*, April 22, 2016. https://www.theguardian.com/cities/2016/apr/22/story-cities-warsaw-rebuilt-18th-century-paintings.

130 *no Marshall Plan for the Pacific:* Constitutional Rights Foundation. https://www.crf-usa.org/bill-of-rights-in-action/bria-20-3-a-the-marshall-plan-for-rebuilding-western-europe.html.

130 *The US did grant some postwar aid:* Agoncillo, *History of the Filipino People*, 450.

130 *to pull on our heartstrings:* This is a poetic interpretation of how laughter increases heart rate variability, and how increased heart rate variability is in turn a measure of decreased autonomic nervous system arousal, i.e., a state of calm (up to a point). For a scientific explanation see Sokichi Sakuragi et al., "Effects of Laughing and Weeping on Mood and Heart Rate Variability." *Journal of PHYSIOLOGICAL ANTHROPOLOGY and Applied Human Science* 21, no. 3 (2002): 159–65. https://doi.org/10.2114/jpa.21.159.

130 *the neocolonial period of Philippine-US relations:* for a journalistic take of this period, see Bulatlat - The Philippines's Alternative Weekly Magazine, https://www.bulatlat.com/news/4-40/4-40-roots.html. For an academic analysis see E. San Juan, Jr., *After Postcolonialism: Remapping Philippines-United States Confrontations* (Lanham: Rowman & Littlefield, 2000).

131 *"We're tired":* Rocky Rivera, quoted in Stephen Bischoff, "Identity and Resistance," in *Empire of Funk: Hip Hop and Representation in Filipina/o America*, ed. Mark R. Villegas, Kuttin' Kandi, and Roderick N. Labrador. (Solana Beach: Cognella, 2014), 247–50.

131 *in order to bring ourselves out of it:* James Baldwin, "The White Man's Guilt," in *Collected* Essays. (New York: Library of America, 1998), 723.

131 *we dig up history, and we work it:* A nod to Missy Elliott, "Work It," track 4 on *Under Construction,* Atlantic Records, 2022.

131 *"goddesses of guerrilla warfare in every lifetime":* A line by Klassy from Ruby Ibarra featuring Rocky Rivera, Klassy and Faith Santilla, "US," track 14 on *CIRCA91,* Beatrock Music, 2017.

131 *great-grandchildren of babaylan who swam with crocodiles:* For more on continuing legacies of the babaylan, see https://www.centerforbabaylanstudies.org/, and Leny Mendoza Strobel, ed. *Babaylan: Filipinos and the Call of the Indigenous* (Santa Rosa: Center for Babaylan Studies, 2010).

131 *The more we root our soles in this history:* As E.J.R. David explained, "Research shows that the more we block out things the more impact they have on our behaviors. For example, research on stereotypical thoughts shows that they [are reinforced] when suppressed. We can apply the same approach to colonialism. There might be some good things but we have to acknowledge it was painful and traumatic. If we don't acknowledge it, it can be debilitating, it can affect our well-being and mental health." E.J.R David, in discussion with the author, September 16, 2019.

III: Neuroregulation

133 *In therapeutic terms:* "About the Journal," *NeuroRegulation*, https://www.neuroregulation.org/about.

133 *maintaining bodily functions:* Though she doesn't use this term in this way, for more on how the brain manages metabolism, heart rate, breathing, and overall "body budget," see Lisa Feldman Barrett, *Seven and a Half Lessons About the Brain* (New York: Houghton Mifflin Harcourt, 2020).

133 *Not even the pain of colonization:* Leny Mendoza Strobel, *Coming Full Circle: The Process of Decolonization Among Post-1965 Filipino Americans* (Santa Rosa, CA: The Center for Babaylan Studies, 2015).

Bayanihan

135 *fight to stay undrowned:* This is a reference to Alexis Pauline Gumbs's incredible work on the lessons she has learned from dolphins, whales, and other marine mammals on survival and living as a Black feminist. See Alexis Pauline Gumbs, *Undrowned: Black Feminist Lessons from Marine Mammals* (Oakland, CA: AK Press, 2020).

135 *Operation Enduring Freedom:* Institute of Medicine (US) Committee on the Initial Assessment of Readjustment Needs of Military Personnel, Veterans, and Their Families. "2: Operation Enduring Freedom and Operation Iraqi Freedom: Demographics and Impact" in *Returning Home from Iraq and Afghanistan: Preliminary Assessment of Readjustment Needs of Veterans, Service Members, and Their Families* (Washington, DC: National Academies Press [US], 2010), https://www.ncbi.nlm.nih.gov/books/NBK220068/.

135 *sending troops to the Southern Philippines:* Larry Niksch, "Abu Sayyaf: Target of Philippine-U.S. Anti-Terrorism Cooperation," *Congressional Research Service Report for Congress*, 12, https://irp.fas.org/crs/RL31265.pdf.

136 *relying on a Post-it note:* Through conversations with friends, I've learned that Post-it therapy is a common thing, at least in my circles.

140 *stigmatized and poorly understood:* For more on stigma and how it can silence us, see "Let's Face It, No One Wants to Talk About Mental Health," McLean Hospital, April 28, 2022, https://www.mcleanhospital.org/essential/lets-face-it-no-one-wants-talk-about-mental-health.

140 *Filipino martial art of Kali:* For more on Kali, see Robert Rousseau, "History of the Martial Art Style of Kali," *LiveAbout*, last modified April 3, 2018, https://www.liveabout.com/history-and-style-guide-of-kali-2308268. For a poetic treatment of Kali, see Michelle Bautista, *Kali's Blade*, (San Francisco: Meritage Press, 2007).

140 *"smiling depression":* Kevin L. Nadal, *Filipino American Psychology: A Handbook of Theory, Research, and Clinical Practice*, 2nd ed. (Hoboken, NJ: John Wily & Sons, 2021).

140 *Centers for Disease Control study:* Quoted in Joyce R. Javier, "The use of an educational video to increase suicide awareness and enrollment in parenting interventions among Filipinos," *Asian American Journal of Psychology* 9, no. 4 (2018), 327–33, doi:10.1037/aap0000144. Note that this study is dated and with a limited sample size, many more studies on Filipina/o/x suicidality and mental health are needed.

141 *pressures to keep group harmony:* Nadal, *Filipino American Psychology*, 30.

141 *Homing Instinct:* "Homing," as a verb describing second-generation immigrants' search for home, is a concept I learned from writer and *The Seventh Wave* founder, Joyce Chen.

141 *bahay kubos:* Aimee Buccuan, "Nipa Hut: The National House of the Philippines," *Alumniyat*, October 26, 2020, https://alumniyat.com/2020/10/26/nipa-hut-the-national-house-of-the-philippines/.

141 *"The Three Little Pigs":* Oliver Tearle, "A Summary and Analysis of the 'Three Little Pigs' Fairy Tale," *Interesting Literature*, https://interestingliterature.com/2020/04/three-little-pigs-fairy-tale-fable-summary-analysis-origins/.

142 *"healing crisis":* Cheryl Deroin, "The Healing Crisis," *Naturopathic Doctor News & Review*, May 20, 2007, https://ndnr.com/autoimmuneallergy-medicine/the-healing-crisis/.

143 *Rituals are often designed:* Francesca Gino and Michael I. Norton, "Why Rituals Work." *Scientific American*, May 14, 2013, https://www.scientificamerican.com/article/why-rituals-work/.

144 *re-walking of the Trail of Tears:* McKenna Princing, "Trail of Tears hike: Bug bites, meditation, maybe life change," *UW Medicine | Newsroom*, May 27, 2014, https://newsroom.uw.edu/story/trail-tears-hike-bug-bites-meditation-maybe-life-change.

144 *annual remembrance:* Re-membering through ritual may be one way to rewire the past and the negative aspects of its hold on our bodies.

Bryan Stevenson, executive director of the Equal Justice Society, was a leading force behind creating the National Memorial for Peace and Justice, a museum dedicated to the memory of lynching victims. See Jonathan Capeheart, "Bryan Stevenson wants us to confront our country's racial terrorism and then say, 'Never again,'" *Washington Post*, April 24, 2018, https://www.washingtonpost.com/blogs/post-partisan/wp/2018/04/24/bryan-stevenson-wants-us-to-confront-our-countrys-racial-terrorism-and-then-say-never-again/.

Palestinians commemorate the 1948 Nakba, in which hundreds of thousands of Palestinians were expelled and religious sites and villages destroyed to create the state of Israel. Without institutional means to create memorials like the National Memorial for Peace and Justice, Palestinians ritualize commemoration through grassroots means: oral history telling, summer camps, and visits to destroyed villages and villages of origin. See Hussein Ibish, "A 'Catastrophe' That Defines Palestinian Identity," *The Atlantic*, May 14, 2018, https://www.theatlantic.com/international/archive/2018/05/the-meaning-of-nakba-israel-palestine-1948-gaza/560294/.

144 *550 miles at gunpoint:* "Trail of Tears," Choctaw Nation. https://edu497 choctawnation.wordpress.com/notable-events/trail-of-tears/.

144 *landmarks along the way:* For more on the remembering of the Trail of Tears and healing historical trauma in practice, see the work of Karina Walters: Walters, "UW-Social Work-Leading Lights Series-Dr. Karina Walters-2021-02-25-330pm Pacific Time."

144 *Fighting Fred:* A satirical tribute to Frederick Funston by Mark Twain: Mark Twain, "A Defense of General Funston," FoundSF, https://www.foundsf .org/index.php?title=A_Defense_of_General_Funston.

145 *modern-day rituals:* From a neuroscience perspective, rituals like these can heal on many levels. By keeping a narrative of these mass traumas present in the explicit memories of diasporic communities and nations, rituals can heal through the creation of a common story-based identity rooted in resilience. Ritual storytelling can also retrain neural circuits by strengthening circuits responsible for creativity, which can eventually lead to these creative circuits activating more often than neural circuits responsible for defeating thoughts. On the level of bodily healing, ritual may act as a method of completing what Peter Levine describes as the incomplete action of a traumatic event. By walking the Trail of Tears and discussing Choctaw futures, the practice may activate the physiology of choosing to walk toward a new home to counteract a freeze response from forced relocation. Both walking (or ambulating in a wheelchair in the case of disabled people) and the act of imagining can help break patterns of defensive stress response caused by the traumatic event. These practices can lead to a sense of home within our own bodies and a stronger sense of agency even in the face of ongoing displacement and duress. Analysis based on the work of Peter A. Levine, *In an Unspoken Voice: How the Body Releases Trauma and Restores Goodness* (Berkeley, CA: North Atlantic Books, 2010).

145 *"colonial trauma response":* Karina L. Walters et al., "Bodies Don't Just Tell Stories, They Tell Histories: Embodiment of Historical Trauma among American Indians and Alaska Natives," *Du Bois Review: Social Science Research on Race* 8, no. 1 (2011), 179–89, doi:10.1017/S1742058X11000 18X.

145 *Filipinos experience a similar colonial trauma:* Strobel, *Coming Full Circle*; Nadal, *Filipino American Psychology*; and David, *Brown Skin, White Minds*.

146 *"bayanihan":* "Bayanihan, house-moving tradition of the Philippines," *Voice of Vietnam*, July 19, 2017, https://vovworld.vn/en-US/content/NDgzMTk0.vov.

146 *the Committee for Human Rights in the Philippines:* Has grown into the International Coalition for Human Rights in the Philippines: https://ichrp.net/.

147 *a strong sense of ethnic identity:* Tiffany Yip, Gilbert C. Gee, and David T. Takeuchi, "Racial Discrimination and Psychological Distress: The Impact of Ethnic Identity and Age Among Immigrant and United States–Born Asian Adults," *Developmental Psychology* 44, no. 3 (2008), 787–800, doi:10 .1037/0012-1649.44.3.787.

147 *voting, volunteering, and activism*: "Voting Counts as a Healthy Habit," *WebMD*, https://www.webmd.com/mental-health/features/voting-counts-as-healthy-habit.

147 *PEP:* For more on Pin@y Educational Partnerships, see http://www.pepsf.org/.

148 *saved her life:* PEP founder Allyson Tintiangco-Cubales on how ethnic studies saves lives: Allyson Tintiangco-Cubales, "Ethnic Studies: Saving Lives, Sacred Spaces & Solidarity | TFCU Talks San Jose: Dr. Allyson," Kollective Hustle, posted on December 19, 2017, YouTube video, https://www.youtube.com/watch?v=2gbXr4diMjU.

148 *a community of value:* Alexis David, in discussion with the author, October 1, 2021. According to Joyce Javier, pride in one's ethnic and cultural heritage is associated with more positive mental health. As told to Agnes Constante, "How 'hiya,' 'kapwa' and other cultural values play a role in Filipino American mental health," *Los Angeles Times*, March 17, 2022, https://www.latimes.com/lifestyle/story/2022-03-17/how-hiya-kapwa-and-other-cultural-values-play-a-role-in-filipino-american-mental-health.

149 *"spoken truth to the lie that I am alone":* Imee Dalton, in discussion with the author, June 6, 2022.

150 *serious mental health issues:* "Living Well with Serious Mental Illness," Substance Abuse and Mental Health Services Administration, last modified June 28, 2022, https://www.samhsa.gov/serious-mental-illness.

150 *electroconvulsive therapy:* This is essentially a therapy to induce epigenetic modifications to neurons. "Electroconvulsion stimulation is used to treat major depression when drugs don't work. It activates many neurons at the same time to induce DNA methylation." Song, discussion.

150 *mental health challenges were un-Filipino:* Interview with Marina (not her real name), October 19, 2021.

150 *disproportionate rates of suicidal ideation:* Alexis Aguilan David, "Building a Home for Filipinx Mental Wellness" in *Closer to Liberation: Pin[a/x]y Activism in Theory and Practice*, ed. Amanda Solomon Aromao, D. J. Kuttin Kandi, and Jen Soriano (Solana Beach, CA: Cognella, 2022).

150 *culturally competent services:* These can include language therapy but also feature therapists who understand both cultural and historical factors that can affect mental health—for example, colonial mentality. See E. J. R. David and Sumie Okazaki, "Colonial mentality: a review and recommendation for Filipino American psychology," *Cultural Diversity & Ethnic Minority Psychology* 12, no. 1 (2006), 1–16, doi:10.1037/1099-9809.12.1.1.

150 *Fil-Ams throughout the United States:* For more on the Filipino Mental Health Initiative of San Francisco, see https://www.fmhi-sf.org/.

151 *antidote to American individualism:* Kevin Nadal has pointed out that there can be a harmful side to collectivism as well when it comes to Filipino American mental health. The Filipino value of pakikisama, or social harmony, could lead some to internalize problems because they don't want to disrupt

group cohesion. As told to Constante, "How 'hiya,' 'kapwa' and other cultural values play a role in Filipino American mental health."

Unbroken Water

152 *I wish I were like a river:* "Danum by Salidummay," Salidummay DKK, posted on August 15, 2010, YouTube video, https://www.youtube.com/watch?v=Q_dpZ3a87qXE.

152 *ground zero in a global battle:* CPA Philippines, "The Popular Resistance to Chico and Cellophil as Self-Determination," *Cordillera Peoples Alliance Posts* (blog), September 8, 2016, https://cpaphils.wordpress.com/2016/09/08/the-popular-resistance-to-chico-and-cellophil-as-self-determination/.

153 *the Missouri River:* "Missouri River," American Rivers, https://www.americanrivers.org/river/missouri-river/.

153 *multiple tribes:* See, for example, "Missouri River," Sacred Land Film Project, https://sacredland.org/missouri-river-united-states/.

153 *largest hydropower project in Asia:* "Marcos to Proceed With River Project Opposed by Tribes," *New York Times*, February 6, 1975, https://www.nytimes.com/1975/02/06/archives/marcos-to-proceed-with-river-project-opposed-by-tribes.html.

153 *1.3 million indigenous people:* In the Cordilleras, indigenous people still live on their land. Spanish colonizers were never able to conquer the Cordillera people, and Americans opted for cultural and economic influence rather than full-on settler colonialism in the region.

Chingmak Phutoli Shikhu, "International Law and Reparations for Indigenous People in Asia" in *Reparations for Indigenous Peoples: International and Comparative Perspectives*, ed. Federico Lenzerini (New York: Oxford University Press, 2008).

153 *Kalinga, Bontoc, and Kankanaey:* For resources by indigenous people living in the US on indigenous culture in the Philippines and the diaspora, see https://www.ikatvoices.com/.

153 *rely on water:* "The Cordillera: Its Land and People," Cordillera Peoples Alliance, last modified December 12, 2006, https://www.cpaphils.org/cordillera.htm.

153 *communal irrigation systems:* Inday Espina-Varona, "Veterans of Chico River Dam struggle join grandkids' generation in new fight," *ABS-CBN News*, April 23, 2018, https://news.abs-cbn.com/focus/04/23/18/veterans-of-chico-river-dam-struggle-join-grandkids-generation-in-new-fight.

153 *sustenance for the native forests:* Karlston Lapniten, "'The river will bleed red': Indigenous Filipinos face down dam projects," *Mongabay*, February 26, 2021, https://news.mongabay.com/2021/02/the-river-will-bleed-red-indigenous-filipinos-face-down-dam-projects/.

The river provides sustenance to animal bodies as well. There are twenty-eight species of wildlife endemic to this specific region.

153 *one hundred thousand:* Lapniten, "'The river will bleed red.'"

The project would have submerged barangay villages in eight towns, displaced one hundred thousand Kalinga and Bontoc people, and impacted at least three hundred thousand people.

153 *sacrifice for national development:* Indigenous people in the Northern and Southern Philippines have been systematically marginalized from and targeted by non-indigenous Philippine society and government. For more on this, see *"ILO in Indigenous and Tribal Peoples in the Philippines,"* International Labour Organization, https://www.ilo.org/manila/areasofwork/WCMS_402361/lang--en/index.htm. Someone once said to me the Philippines was never meant to be a nation. It is made of hundreds of different peoples speaking more than 150 different languages, forced into nationhood by Spanish colonial rule. The indigenous people of the Cordilleras and the indigenous people of Mindanao do not necessarily consider themselves Filipinos. There are 110 ethnolinguistic groups that make up indigenous people and cultural minorities (about fourteen million people or 20–25 percent of the Filipino population as a whole). They are Bontoc, Ifugao, Kankanaey, Manobo, T'boli, Tasaday; they are their ethnolinguistic tribal identities. Joseph Ruanto Ramirez, "IgoROOTS and Allyship," IKAT Collective webinar, August 2020, https://www.ikatvoices.com.

154 *"black snake" pipeline:* Nick Estes and Jaskiran Dhillon, ed. *Standing with Standing Rock: Voices from the #NoDAPL Movement*, (Minneapolis, MN: University of Minnesota Press, 2019).

154 *lusay:* Sofia K. Guanzon, "This comic documents the role of women in the Cordillera people's movement," *CNN Philippines*, December 10, 2021, https://www.cnnphilippines.com/life/culture/literature/2021/12/10/This-comic-documents-the-role-of-women-in-the-Cordillera-peoples-movement.html.

154 *police headquarters in Bolinao:* Leticia Bula-at, "*INDIGENOUS WOMEN'S STRUGGLES: The Chico Dam Project and the Kalinga Women*," trans. Bernice See (paper presented at the NGO Forum of the Fourth World Conference on Women 1995, Forum on Women's Resistance Strategies), https://cpcabrisbane.org/Kasama/1996/V10n2/Innabuyog.htm.

154 *basketballs and chocolate:* Cordillera People's Alliance, "The Cordillera Resistance Against Chico Dam and Cellophil," Bantayog Foundation, October 10, 2015, https://bantayog.foundation/the-cordillera-resistance-against-chico-dam-and-cellophil.

154 "*I do not have anything to sell*"*:* Jerrie Abella, "Indigenous People Remember Martyr Macliing Dulag's martyrdom," *GMA News*, April 24, 2010, https://www.gmanetwork.com/news/topstories/nation/189239/indigenous-people-remember-macliing-dulag-s-martyrdom/story/.

154 *Bodong Federation:* Laurence L. Delina, "Indigenous environmental defenders and the legacy of Macli-ing Dulag: Anti-dam dissent, assassinations, and protests in the making of Philippine energyscape," *Energy Research & Social Science* 65 (2020), 101463, doi:10.1016/j.erss.2020.101463.

They sealed the alliance through a multilateral peace pact, or pagta, consolidating widespread regional resistance to the dams and also attracting support from churches, academics, NGOs, and environmental organizations from across the globe.

155 *"free-fire zones":* Delina, "Indigenous environmental defenders and the legacy of Macli-ing Dulag."

155 *Dulag died instantly:* "Remembering Macli-ing Dulag and the Anti-Chico Dam Struggle," Human Rights Violations Victims' Memorial Commission, April 24, 2021, https://hrvvmemcom.gov.ph/remembering-macli-ing-dulag-and-the-anti-chico-dam-struggle/.

155 *"nourished by our blood":* Lapniten, "'The river will bleed red.'"

155 *successful struggle:* "'The river will bleed red': Indigenous Filipinos face down dam projects," Mongabay, posted on February 26, 2021, YouTube video, https://www.youtube.com/watch?v=TFfyZ07YNhI.

155 *1997 Philippine Indigenous Peoples' Rights Act:* "Free, Prior and Informed Consent," Food and Agriculture Organization of the United Nations, https://www.fao.org/indigenous-peoples/our-pillars/fpic/en/.

157 *one of the four:* Frank Cimatu, "Locals protest Duterte admin's Chico River project," *Rappler*, June 24, 2017, https://www.rappler.com/nation/173866-protest-duterte-admin-chico-river-project/. Duterte also attempted to build a dam in Dupag but was met with effective opposition.

157 *palay stretches to the horizon:* Rice is a staple crop of the Cordilleras. More recently, there have ben efforts to preserve and expand distribution of heirloom rice varieties. See Subir Bairagi et al., "Preserving cultural heritage through the valorization of Cordillera heirloom rice in the Philippines," *Agriculture and Human Values* 38, no. 1 (2021), 257–70, doi:10.1007/s10460-020-10159-w.

159 *like it is literally alive:* While living by the banks of the Chico River, I kept thinking of Leny Mendoza Strobel's quote: "They say we are a people who lived in a convent for three hundred years and fifty years in Hollywood. . . . But underneath this veneer, the ancient spirits never died, the anitos never slept." Strobel, *Coming Full Circle*. In some precolonial Philippine beliefs, water is not only life; it is alive, and it is also the bridge to the next life. The other world is believed to lie beyond a wide river in the realm where the sun drowns. Potet, *Ancient Beliefs and Customs of the Tagalogs*.

159 *have fought to keep it that way:* For a history of armed conflict in the Cordillera region, see Miriam Coronel Ferrer, "4. The Moro and the Cordillera Conflicts in the Philippines and the Struggle for Autonomy," in *Ethnic Conflicts in Southeast Asia*, ed. Kusuma Snitwongse and W. Scott Thompson (Singapore: ISEAS Publishing, 2005), 109–50, doi: 10.1355/9789812305565-007.

161 *pinapaitan:* Raymund, "Pinapaitan," *Ang Sarap* (blog), August 13, 2013, https://www.angsarap.net/2013/08/13/pinapaitan/.

161 *how they steward:* Leni D. Camacho et al., "Indigenous knowledge and

practices for the sustainable management of Ifugao forests in Cordillera, Philippines," *International Journal of Biodiversity Science, Ecosystem Services & Management* 12, no. 1–2 (2016), 5–13, doi:10.1080/21513732.2015.1124453.

161 *This system:* Camacho et al., "Indigenous Knowledge and Practices for the Sustainable Management of Ifugao Forests in Cordillera, Philippines."

161 *Sapi Bawer:* Nick Street, "Sapi's Struggle: Indigenous Resistance to Cultural Assimilation in the Philippines," USC Dornsife Center for Religion and Civic Culture, January 27, 2021, https://crcc.usc.edu/sapis-struggle-indigenous-resistance-to-cultural-assimilation-in-the-philippines/.

162 *pattung:* While at Cordillera Day, we learned the pattung style where you play the gangsas standing and hold them in your hands, keeping step with dancers and beating the gangsas with padded wooden sticks. See "Cordillera's Musical Instruments," MUSIKO CORDILLERA, http://musikocordillera.weebly.com/musical-instruments.html.

162 *San Roque Dam:* "San Roque Dam," International Rivers, https://archive.internationalrivers.org/campaigns/san-roque-dam.

163 *"afraid they will get shot too":* Francis Dulag, in discussion with the author, April 24, 2002. Francis is Macli-ing Dulag's eldest son and also goes by Loc-an.

164 *Cordillera legend:* F. Landa Jocano, *Outline of Philippine Mythology* (Manila, Philippines: Centro Escolar University Research and Development Center, 1969), 109.

164 *one of the most common photos:* See, for example, Hanah Tabios, "International Civil Society, Human Rights Groups Defend Macli-Ing Dulag's Monument from Demolition," *Manila Bulletin*, October 19, 2020, https://mb.com.ph/2020/10/19/international-civil-society-human-rights-groups-defend-macli-ing-dulags-monument-from-demolition/.

164 *"build, build, build":* Aika Rey, "Build, Build, Build: Mapping the Duterte administration's infrastructure legacy," *Rappler*, June 28, 2022, https://www.rappler.com/business/build-build-build-mapping-duterte-administration-infrastructure-legacy/.

164 *the fate of the Chico:* Daniza Fernandez, "Bongbong Marcos to continue Duterte's infrastructure program," *Philippine Daily Inquirer News*, July 25, 2022, https://newsinfo.inquirer.net/1634242/marcos-to-continue-dutertes-infra-program.

164 *at the expense of those downstream:* Lapniten, "'The river will bleed red.'"

164 *self-determination will prevail:* Delina, "Indigenous environmental defenders and the legacy of Macli-ing Dulag." As Delina writes, "As long as these development projects fail to embed concepts of justice in the making of an energyscape, Indigenous Peoples' struggles to speak truth to power are also poised to continue."

165 *humans and nature:* Carolyn Merchant, *The Death of Nature: Women, Ecology, and the Scientific Revolution* (San Francisco: HarperOne, 1990).

The Western division of humans and nature has its origins in the Scientific Revolution, in which the earth was seen as more of a machine to be leveraged for industrial gain than an organic entity to be stewarded. This change led to gendered concepts of dominating nature that influenced gendered concepts of the domination and control of women.

165 *revolution of values:* from Martin Luther King's 1967 "Beyond Vietnam" speech. See "A Revolution of Values," Zinn Education Project, https://www.zinnedproject.org/materials/revolution-of-values/.

165 *his grandchildren have joined him in the ritual:* Espina-Varona, "Veterans of Chico River Dam struggle join grandkids' generation in new fight."

165 *Nature functions as an autonomic nervous system regulator:* V. F. Gladwell et al., "The effects of views of nature on autonomic control," *European Journal of Applied Physiology* 112, no. 9 (2012), 3379–86, doi:10.1007/s00421-012-2318-8. Nature activates our parasympathetic states of rest, which can help us heal. Patients who undergo surgery recover faster when they can see trees through a window compared to patients who have to look at a brick wall. See Jim Robbins,"Ecopsychology: How Immersion in Nature Benefits Your Health," *Yale Environment 360*, January 9, 2020, https://e360.yale.edu/features/ecopsychology-how-immersion-in-nature-benefits-your-health.

Studies have shown that spending at least two hours per week in nature increases people's well-being, from more positive moods to strengthened immune function, lower blood pressure, and lower amounts of stress hormones.

166 *Bathala:* Bathala was ungendered; I'm taking poetic license to call her a goddess. Potet, *Ancient Beliefs and Customs of the Tagalogs.*

Awaken the Lyrics

169 *Awaken the Lyrics:* From Kimmortal, "Stars," track 1 on *X marks the Swirl*, Coax Records, 2019.

169 *Diskarte Namin:* "Rise Up," "Diaspora," "100 Years," and "River" lyrics: Diskarte Namin, creative commons copyright 2010-noncommercial, attribution, share-alike. https://creativecommons.org/licenses/by-nc-sa/4.0/.

169 *"superpredator":* "The Superpredator Myth, 25 Years Later," Equal Justice Initiative, April 7, 2014, https://eji.org/news/superpredator-myth-20-years-later/.

170 *Proposition 21:* Elizabeth Martinez, "The New Youth Movement In California," *Third World Traveler* (blog), May 2000, https://thirdworldtraveler.com/Children/NewYouthMovement_Calif.html.

170 *getting pushed out of the city:* James Tracy, "A Decade of Displacement," FoundSF. https://www.foundsf.org/index.php?title=A_Decade_of_Displacement.

170 *The sound of the Bay in the early 2000s:* Alan Chazaro, "'We Were Hyphy' and the Glory of Mid-2000s Bay Area Hip-Hop," *The San Francisco Standard*, April 2, 2022, https://sfstandard.com/arts-culture/we-were-hyphy-revisits-the-glory-days-of-mid-2000s-bay-area-hip-hop/.

"Zion I," Wikipedia, last modified January 26, 2022, https://en.wikipedia.org/w/index.php?title=Zion_I&oldid=1068132094.

"The Coup: Party Music," *Pitchfork*, January 3, 2002, https://pitchfork.com/reviews/albums/1633-party-music/.

170 *Local 1200:* DJ Sake 1, "(((Local 1200))) 10 YEARS + COUNTING...," August 1, 2007, http://jediknightlife.blogspot.com/2007/08/local-1200-10-years-counting.html.

170 *"Another One Bites the Dust.":* Queen, "Another One Bites the Dust," track 3 on *The Game*, EMI, 1980.

170 *the Cell:* Russell Howze, Jonathan Youtt, and Devin Holt, "CELLspace: 1996-2012," FoundSF, https://www.foundsf.org/index.php?title=CELLspace:_1996-2012.

172 *anhedonia:* Oliver Sacks, *Musicophilia: Tales of Music and the Brain* (New York: Vintage, 2008).

173 *"Real Love":* Mary J. Blige, "Real Love,"track 4 on *What's the 411?*, Uptown Records, 1992.

173 *"our strategy":* Big M came up with this name, explaining that it was what people said in the Philippines when playing cards: "Ano ang disakarte mo?" meaning, "What's your game plan; what's your strategy to win?"

174 *collective culture of dissent:* Sandra Lynn Curtis, "Music Therapy and Social Justice: A Personal Journey," *The Arts in Psychotherapy* 39, no. 3 (2012), 209–13, doi:10.1016/j.aip.2011.12.004.

174 *"This is who we are":* Christopher Small, *Musicking: The Meanings of Performing and Listening* (Middletown, CT: Wesleyan University Press, 1998), 134.

174 *attention modulator:* Stefan Koelsch, "A Neuroscientific Perspective on Music Therapy," *Annals of the New York Academy of Sciences* 1169, no. 1 (2009), 374–84, doi:10.1111/j.1749-6632.2009.04592.x.

174 *negative sensations and emotions:* Music may have also helped me increase what therapists call "the zone of tolerance around trauma" by contributing to a buffer of emotional resilience.

175 *make money overseas:* In 2019, there were more than 2.2 million Filipinos working overseas. For more on OFWs, see Robyn Magalit Rodriguez, *Migrants for Export: How the Philippine State Brokers Labor to the World* (Minneapolis, MN: University of Minnesota Press, 2010); and Jessie Yeung and Xyza Cruz Bacani, "When love is not enough," *CNN*, https://www.cnn.com/interactive/2020/11/asia/hong-kong-filipino-helpers-dst/.

175 *trafficked into the sex trade:* Thousands of Filipinas become victims of human trafficking when they leave home to work overseas: "Overseas Filipino workers vulnerable to trafficking will be protected under new Philippines Department of Migrant Workers," ASEAN–Australia Counter Trafficking, February 2, 2022, https://www.aseanact.org/story/ople/.

176 *Whenever I sang "Diaspora":* "Singing uses all muscles of social engagement system – exhaling slowly has a vagal effect on the heart. It's calming and brings people to you." Porges, discussion.

176 *entranced presence:* Darin McFadyen, "Neuroscience shows listening to music has kind of the same effect as meditation," *Quartz*, May 12, 2018, https://qz.com/quartzy/1274667/neuroscience-shows-listening-to-music-has-kind-of-the-same-effect-as-meditation/.

177 *visited by the FBI:* Farah Pandith, "The U.S., Muslims, and a Turbulent Post-9/11 World," Council on Foreign Relations, September 1, 2021, https://www.cfr.org/article/us-muslims-and-turbulent-post-911-world.

178 *the United States bought the Philippines:* "Introduction - The World of 1898: The Spanish-American War," Hispanic Division Library of Congress, https://loc.gov/rr/hispanic/1898/intro.html.

178 *there must be a lyre:* Judith Kitchen, "Mending Wall," *Seneca Review* XXXVII, no. 2 (2007), 45–49.

179 *"storying back" against colonial erasure:* Bobis, "Decolonial Poetics: Panel with Merlinda Bobis and Rick Barot."

179 *"Paid in Full":* Eric B. & Rakim, "Paid in Full," track 6 on *Paid in Full*, Island Records and 4th & B'way, 1987.

180 *singing in Tagalog:* Paul Bolick-Mausisa, *Kultural Guerrillas: A Comparative Study of Filipino and Chicano/Mexicano Communities through Music and Poetry* (Riga, Latvia: VDM Verlag, 2009), 129.

180 *a people's version of history through song:* As Diskarte member Paul Bolick-Mausisa writes, "Providing 'correct' information about a community's historical and current events and issues is a central task for community organizers and political activists." Bolick-Mausisa, *Kultural Guerrillas*, 108.

181 *like the people's movement to be free:* "Danum by Salidummay," Salidummay DKK.

181 *8th Wonder:* "8TH WONDER // The Visit // Eternal.Infinite," Manalo Pictures, posted on October 20, 2016, Vimeo video, https://vimeo.com/188091705.

181 *Filipinos for Global Justice, Not War:* Irene Duller, "Filipino Coalition for Global Justice Not War Joined April 20 Protest In San Francisco," Indybay, April 24, 2002, https://www.indybay.org/newsitems/2002/04/24/1247151.php.

183 *entrainment in action:* Robert J. Ellis and Julian F. Thayer, "Music and Autonomic Nervous System (Dys)function," *Music Perception* 27, no. 4 (2010), 317–26, doi:10.1525/mp.2010.27.4.317.

183 *"Rhythm binds together":* Sacks, *Musicophilia.*

183 *awakens the reward centers in our brains:* David H. Zald and Robert J. Zatorre, "Chapter 19: Music," in *Neurobiology of Sensation and Reward*, ed. Jay A. Gottfried (Boca Raton, FL: CRC Press, 2011).

184 *greatly undervalued treatment for trauma:* Bruce D. Perry and Oprah Winfrey, *What Happened to You?: Conversations on Trauma, Resilience, and Healing* (New York: Flatiron Books, 2021).

184 *struggled with addiction:* There's not much data on substance use in Filipino American communities, but anecdotal evidence shows that it is high. Elle

de la Cruz says that Filipino cultural values of shame and collectivism, plus the availability of alcohol, tobacco, and other drugs, lead to people turning to them to cope and deal with emotional trauma, rather than turning to others. As told to Agnes Constante, "Talking about substance use can be hard for Filipino Americans. Why it's helpful to share stories," *Los Angeles Times*, September 22, 2022, https://www.latimes.com/lifestyle/story/2022-09-22/filipino-american-mental-health-substance-use-disorders.

184 *Saico:* Doble Bara, https://www.youtube.com/channel/UCjMpLckEu9OkGDYf2_Kc92g.

185 *Aztlan Underground:* Instagram @aztlanunderground

185 *Quetzal:* Instagram @quetzalmusic

185 *Blackfire:* "'Blackfire' Siblings Rock It Out," November 25, 2010, in *Tell Me More*, podcast, MP3 audio, https://www.npr.org/templates/story/story.php?storyId=131565918.

185 *Pepe Manegdeg and Albert Terredano:* Urgent Appeals Programme, "UPDATE (Philippines): Two more activists killed, one of whom suffered twenty-two gunshot wounds," Asian Human Rights Commission, http://www.humanrights.asia/news/urgent-appeals/UP-158-2005/.

185 *resistance and revolutionary love:* Daya Mortel, "'Do You Know How Freedom Sounds': The Chico River Song and Struggles Today" (unpublished paper, University of Hawaii, 2006), digital reproduction of original work.

185 *Sama Sama:* Sama Sama is a cooperative camp in the California Bay Area that develops young Filipinx leaders through activities at the intersection of art, culture, and ecology: https://samasamacooperative.org/.

IV: Neuroplasticity

187 *The capacity of neurons:* Matt Puderbaugh and Prabhu D. Emmady, *Neuroplasticity* (Treasure Island, FL: StatPearls, 2022).

187 "*Underwater, she endures*": Excerpted from Angela Peñaredondo, "Becoming a Minke Whale," *in nature felt but never apprehended* (Blacksburg: Noemi Press, 2023).

Unconditional

190 *"super coregulator":* "Stephen Porges, PhD on Political Differences at Holiday Dinners," NICABM, https://www.nicabm.com/political-differences-at-holiday-dinners/.

190 *comfortable in my own skin:* Trauma therapist Bessel van der Kolk writes that traumatized people feel chronically unsafe in our own bodies. See van der Kolk, *The Body Keeps the Score*. Psychiatrist Judith Herman has shown that reestablishing a sense of safety is the first critical step in trauma recovery. See Herman, *Trauma and Recovery*.

190 *constricted my life:* Psychiatrist Judith Hermann explains that traumatized

people constrict their lives in order to create some sense of safetyHerman, *Trauma and Recovery*, 45.

190 *a group of porcupines:* "Tara Brach: The Courage to Love, Part I," Tara Brach, posted on July 21, 2019, YouTube video, https://www.youtube.com/watch?v=h28cqYP3h2o.

192 *100 percent okay on my own:* Dr. Bruce Perry says that to be engaged with other people, we all need some reciprocal feedback. I gave very little of that feedback, and BJ still wanted to be my friend. Perry and Winfrey, *What Happened to You?*, 58.

192 *Sly Stone playing from your car stereo:* Bill Kopp, "Before the Family Stone: The Viscaynes," *Goldmine*, March 15, 2020, https://www.goldminemag.com/interviews/before-family-stone-viscaynes.

193 *Changó:* "Dance & Santeria | Chango: God of Thunder," Havana Salsa People, https://havanapeoplesalsa.co.uk/cuban-salsa-chango-santeria-dance/.

193 *only a matter of seconds:* "The neural epigenome is extremely dynamic. There are thousands of sides to the genome that can be modified very rapidly through methylation. Many of these modifications are long-lasting. One single activation can change methylation that lasts for days." Hongjun Song (director of neuroepigenetics, University of Pennsylvania), in discussion with the author, November 16, 2020. Perry and Winfrey, *What Happened to You?*, 145.

194 *a new pattern on the kulintang:* For more on the kulintang and the master who taught BJ and brought the tradition the US, see "Danongan Kalanduyan," National Endowment for the Arts, https://www.arts.gov/honors/heritage/danongan-kalanduyan.

194 *dry bears and copy machines:* Pilipino accent jokes as told to me by BJ and Big M from Diskarte: (1) Four bears drive over the Bay Bridge, and their car falls into the water. Which one of them doesn't get wet? The dry bear (driver)! (2) Two Filipino families went camping, and one asked the other why he had a Xerox in the back of his car. Cuz you told me to bring the copy (coffee) machine!

195 *"same wavelength":* Peter Lakatos, Joachim Gross, and Gregor Thut, "A New Unifying Account of the Roles of Neuronal Entrainment," *Current Biology* 29, no. 18 (2019), R890–905, doi:10.1016/j.cub.2019.07.075.

195 *neuroception:* Porges, *The Polyvagal Theory*.

196 *another person's nervous system:* Barrett, *Seven and a Half Lessons About the Brain*.

198 *our hurts can keep us separate:* Ogden, interview. For more about how unresolved trauma can impact relationships and interventions for healing trauma, see the work of Pat Ogden: https://sensorimotorpsychotherapy.org/.

199 *"Reasons":* Earth, Wind & Fire, "Reasons," track 6 on *That's the Way of the World*, Columbia Records, 1975.

199 *one of the most depressing songs ever made:* I recognize that some would fight me on this. It's personal; I associate Reasons with a particular ex—hence, depressing.

199 *turns sickness into terminal disease:* See Nadal, *Filipino American Psychology*; "Maria Yellow Horse Brave Heart: Historical Trauma and Healing in Native Communities"; Harris, *The Deepest Well*; and Joy DeGruy, *Post Traumatic Slave Syndrome: America's Legacy of Enduring Injury and Healing* (self-pub, Joy Degruy Publications, 2017).

200 "*clean" and "dirty" pain:* Menakem, *My Grandmother's Hands.*

200 *memorial fund:* The BJ Alisago Memorial Artistic Scholarship: https://www.bjamscholarship.org/.

Broken Water

201 *before devouring them whole:* Herminia Meñez, *Explorations in Philippine Folklore* (Manila: Ateneo de Manila University Press, 1996), 86–94.

202 *my body might detach:* "Manananggal" comes from the Tagalog word "tang gal," meaning "to detach or separate": See Alex Postrado, "Manananggal: Self-Segmenting, Fetus-Eating Monster," *LoreThrill*, https://lorethrill.com/manananggal-fetus-eating-monster/.

202 *Filipino pop culture: Movies, see for example,* "How horrors like the Manananggal influenced a Filipino American director's bloody new teen vampire film," https://nextshark.com/blumhouse-teen-vampire-film-maritte-go.

Plays: See, for example, "Monstress," Flying V Theater, https://www.flyingvtheatre.com/wings/theatre/fv22-theatre/monstress/.

Books: My favorite recent book that features an aswang, which is a broader category of monster of which the manananggal is one type, is Melissa Chadburn's *A Tiny Upward Shove* (New York: Farrar, Straus and Giroux, 2022).

Trese: Budjette Tan and Kajo Baldisimo, *Trese*. 7 vols. (Portland, OR: Ablaze, 2020). Netflix show: *Trese*, directed by Jay Oliva et al., aired June 10, 2021, https://www.netflix.com/title/81012541.

202 *the foil to the ultimate mother:* Jeanylyn Lopez, "Manananggal: Meet the vampire-like mythical creature of the Philippines," *Medium*, March 9, 2021, https://medium.com/the-hyphenated-filipino/manananggal-meet-the-vampire-like-mythical-creature-of-the-philippines-d676b207aedf.

204 *corrective measures:* Matthew D. Lieberman, *Social: Why Our Brains Are Wired to Connect* (New York: Crown Publishers, 2013), 278.

204 *Chiropractic, acupuncture, spinal traction:* There are no shortage of "alternative" therapies for chronic pain and C-PTSD, and everything I pursued helped me manage pain. But the therapies I found most curative were somatic therapies—specifically focusing, somatic touch, and somatic experiencing—the structural work I did with the Bone Whisperer, yoga, and EMDR.

204 *spent the majority of my income:* The fact that these therapies are mostly out-of-pocket and not covered by insurance is a major barrier to equitable healing from trauma. Americans spend more than thirty billion dollars a year on "complementary and integrative" health treatment not covered by insurance. "Paying for Complementary and Integrative Health Approaches,"

National Center for Complementary and Integrative Health, last modified June 2016, https://www.nccih.nih.gov/health/paying-for-complementary-and-integrative-health-approaches.

204 *a healing sense of agency:* On the importance of agency in healing trauma, see Daniela Rabellino et al., "Altered Sense of Body Ownership and Agency in Posttraumatic Stress Disorder and Its Dissociative Subtype: A Rubber Hand Illusion Study," *Frontiers in Human Neuroscience* 12 (2018), 163, doi:10.3389/fnhum.2018.00163.

205 *attachment wound:* Benamer and White, *Trauma and Attachment.*

205 *also as a healer:* These depictions of the manananggal as a false healer were recounted to me by people in the Philippines and overlap with documented stories of another aswang called Mangangaway, the goddess of sickness and disease. See "ASWANG #71 - Mangangaway," OpenSea, https://opensea.io/assets/ethereum/0x495f947276749ce646f68ac8c248420045cb7b5e/45420177711653843954781376584544552839302508263742502603255103502076017115137.

206 *charted my ovulation cycle:* I learned how to chart my cycle and check for signs of ovulation from the book by Toni Weschler, *Taking Charge of Your Fertility* (New York: William Morrow Paperbacks, 2006).

206 *a variety of tests:* The most excruciating of these was the hysterosalpingogram, or HSG, a test that involves flushing an iodine solution through the fallopian tubes to investigate potential causes of infertility. For more, see Raymond Chieng, "Hysterosalpingogram," *Radiopaedia*, last modified October 5, 2022, https://radiopaedia.org/articles/hysterosalpingogram?lang=us.

206 *straight to IVF:* I am eternally grateful to all those who contributed to the science of IVF. I would not have my child without them. For a brief history of IVF, see "The History of IVF -The Milestones," IVF Worldwide, https://ivf-worldwide.com/ivf-history.html.

207 *for their own good:* The feelings of disgust that I experienced were distinct from the nineteenth-century sense of shame and disgust of pregnancy that was dominant among white America, as chronicled by Richard and Dorothy Wertz in *Lying-In: A History of Childbirth in America* (New Haven, CT: Yale University Press, 1989), 80–81. These sentiments were born of sexual and physical prudery and other dominant patriarchal cultural feelings of pregnancy and sex as dirty. Dirt and prudery were not present in my feeling of disgust. Violation, vulnerability, and the fear of wearing down my defenses, the body's natural weaponry and abilities to fight back, were the feelings that overcame me.

207 *bamboo stake through her back:* "From Aswang to Adobo." *The Official Newsletter of the Horror Writers Association*, http://www.horror.org/private/newsletter/october-2016/195-aswang.html.

208 *ash on her torso:* PBS Digital Studios, "Monstrum: Manananggal," Described and Captioned Media Program, released 2019, video, https://dcmp.org/media/13315-monstrum-manananggal.

208 *she will die before dawn:* "Manananggal." Cryptid Wiki, last modified June 18, 2020, https://cryptidz.fandom.com/wiki/Manananggal.

209 *I could be a Filipinx Edna Pontellier:* Edna Pontellier, the protagonist from Kate Chopin's *The Awakening*, drowns herself in the last scene of the book. The scene of her swimming into deeper and deeper water has always stayed with me.

210 *stronger feelings of disgust:* Christal L. Badour, and Matthew T. Feldner, "The Role of Disgust in Posttraumatic Stress: A Critical Review of the Empirical Literature." *Journal of Experimental Psychopathology* 9, no. 3 (2018), pr.032813, doi:10.5127/pr.032813. This is partly because a feature of PTSD is a dampened ability to regulate emotions, and partly because there is a relationship between PTSD and phobias associated with intense disgust.

210 *undesirable individuals:* Paul Rozin, Laura Lowery, and Rhonda Ebert, "Varieties of Disgust Faces and the Structure of Disgust," *Journal of Personality and Social Psychology* 66, no. 5 (1994), 870–81, doi:10.1037//0022-3514.66.5.870.

210 *a study about tokophobia:* Manjeet Singh Bhatia and Anurag Jhanjee, "Tokophobia: A dread of pregnancy," *Industrial Psychiatry Journal* 21, no. 2 (2012), 158–59, doi:10.4103/0972-6748.119649.

Tokophobia is a fear of pregnancy and birth in those who have never been pregnant. There is also secondary tokophobia, which is a learned fear by people who have had traumatic pregnancies or labor. The one scientific study that has been done on tokophobia found that as many as eight in ten pregnant women reported some level of worry, but that the pathological fear of pregnancy and birth is more rare.

211 *"magtatangal":* "Manananggal." Wikiwand, https://wikiwand.com/en/Mana nanggal.

211 *to diminish the babaylan's power:* Hope Sabanpan-Yu, "Performing the Body in Filipino Narratives: The Manananggal (Viscera Sucker) in Colonial Literature," *UNITAS* 90, no. 1 (2017), 57–73, doi: 10.31944/2017901.HO-SAYU03.

212 *The polyvagal theory:* Porges, *The Polyvagal Theory*.

212 *inhibited in healthy individuals:* Psychiatrist Louis Cozolino explains, "A vital aspect of the development of the cortex is inhibitory—first of reflexes, later of spontaneous movements and even later of emotions and inappropriate social behavior." Cozolino, *The Neuroscience of Psychotherapy*, 69.

212 *impaired due to age:* Cozolino, *The Neuroscience of Psychotherapy*, 70.

212 *The vagal brake:* Porges, *The Polyvagal Theory*. In healthy adults, the more recently evolved circuits responsible for social interaction inhibit defense circuits. This is what is known as the vagal brake in polyvagal theory. Many neuroscientists, including some whom I interviewed for this book, dislike the polyvagal theory because of the lack of rigorous scientific evidence for the theory. But it has been an extremely useful model for therapists, such as those involved with the National Institute for the Clinical Application of Behavioral Medicine. See https://www.nicabm.com/.

213 *craniosacral sessions in a saltwater pool:* KinFloat is an incredible water therapy program in the Río Piedras neighborhood of San Juan, Puerto Rico. See https://www.terapiasfloat.com/.

213 *Relaxin:* For more on the wonders of relaxin, see F. Dehghan et al., "The effect of relaxin on the musculoskeletal system," *Scandinavian Journal of Medicine & Science in Sports* 24, no. 4 (2014), e220–29, doi:10.1111/sms.12149.

213 *an elder manananggal:* PBS Digital Studios, "Monstrum: Manananggal."

213 *eating away at her entrails:* "Philippine Folklore: Meet the Vampiric, Cannibalistic, Manananggal," Under the Influence, May 17, 2017, https://zteve tevans.wordpress.com/2017/05/17/philippine-folklore-meet-the-vampiric -cannibalistic-manananggal/.

214 *highest maternal:* US maternal mortality rates: Roosa Tikkanen et al., "Maternal Mortality and Maternity Care in the United States Compared to 10 Other Developed Countries," The Commonwealth Fund, November 18, 2020, https://www.commonwealthfund.org/publications/issue-briefs/2020 /nov/maternal-mortality-maternity-care-us-compared-10-countries. US infant mortality rates: Rabah Kamal, Julie Hudman, and Daniel McDermott, "What do we know about infant mortality in the U.S. and comparable countries?," Peterson-KFF Health System Tracker, October 18, 2019, https://www .healthsystemtracker.org/chart-collection/infant-mortality-u-s-compare -countries/.

214 *infant mortality rates:* "2019 Annual Report: International Comparison," America's Health Rankings, https://www.americashealthrankings.org/learn/ reports/2019-annual-report/international-comparison.

214 *two to three times the rate of maternal and infant death:* Hill, Latoya, et al. "Racial Disparities in Maternal and Infant Health: Current Status and Efforts to Address Them." *KFF*, 1 Nov. 2022, https://www.kff.org/racial-equity-and -health-policy/issue-brief/racial-disparities-in-maternal-and-infant-health -current-status-and-efforts-to-address-them/.

214 *disabled people:* Paige Falion and Mallory Cyr, "Disability & Reproductive Health: It's Time to Talk About It!" AMCHP, March 2022, https://amchp .org/2022/03/17/disability-reproductive-health-its-time-to-talk-about-it/.

214 *queer and trans people:* Ruth Dawson and Tracy Leong, "Not Up for Debate: LGBTQ People Need and Deserve Tailored Sexual and Reproductive Health Care," Guttmacher Institute, November 16, 2020, https://www.gutt macher.org/article/2020/11/not-debate-lgbtq-people-need-and-deserve -tailored-sexual-and-reproductive-health.

214 *affected by many other issues:* Low birth weight can have consequences later in life and lead to disparities in adult disease. See Christopher W. Kuzawa and Elizabeth Sweet, "Epigenetics and the Embodiment of Race: Developmental Origins of US Racial Disparities in Cardiovascular Health," *American Journal of Human Biology* 21, no. 1 (2009), 2–15, doi:10.1002/ajhb.20822.

214 *Puerto Rico's colonial economy:* Lara Merling, "Puerto Rico's Colonial Legacy and Its Continuing Economic Troubles," Center for Economic and Policy

Research, September 20, 2018, https://cepr.net/puerto-rico-s-colonial-legacy-and-its-continuing-economic-troubles/.

215 *trauma-informed birthing:* See https://www.pennysimkin.com/.

216 *eat real food:* On eating during labor: Colleen de Bellefonds, "Can You Eat or Drink During Labor?," What to Expect, May 4, 2022, https://www.whattoexpect.com/pregnancy/eating-well/eating-during-labor/.

217 *placental separation:* Placental complications have been found to be more common in IVF and other types of technology-assisted pregnancies. See UNSW Media, "ART pregnancies at higher risk of placental complications," UNSW Newsroom, May 18, 2018, https://newsroom.unsw.edu.au/news/health/art-pregnancies-higher-risk-placental-complications.

218 *"Don't stop believing, hold on to that feeling":* Journey, "Don't Stop Believin'," track 1 on *Escape*, Columbia Records, 1981.

V: Neuromimicry

221 *drawing inspiration from nervous system design:* I made up this definition, inspired by biomimicry at large. See "What Is Biomimicry," Biomimicry Institue, https://biomimicry.org/what-is-biomimicry/.

221 *a ghost hums through my bones:* Yusef Komunyakaa, "Anodyne" in *Pleasure Dome: New and Collected Poems* (Wesleyan University Press, 2001).

Rupture

223 *how to regulate themselves:* Andrea C. Buhler-Wassmann and Leah C. Hibel, "Studying caregiver-infant co-regulation in dynamic, diverse cultural contexts: A call to action," *Infant Behavior and Development* 64 (2021), 101586, doi:10.1016/j.infbeh.2021.101586.

223 *"You Are the Sunshine of My Life":* Stevie Wonder, "*You Are the Sunshine of My Life*," track 1 on *Talking Book*, Tamla, 1972.

224 *"What an incredible place to be":* Eduardo Duran, *Healing the Soul Wound: Trauma-Informed Counseling for Indigenous Communities*, 2nd ed. (New York: Teachers College Press, 2019).

224 *Living with trauma-related anxiety, C-PTSD:* For more on C-PTSD and its effects, see Foo, *What My Bones Know*.

224 *illness metaphor of spoons:* Christine Miserandino, "The Spoon Theory," But YouDontLookSick.com, https://butyoudontlooksick.com/articles/written-by-christine/the-spoon-theory/.

225 *a state of slash-and-burn:* While parenting and writing this book, I couldn't find much in the way of literature online or IRL that discussed parenting with C-PTSD. The most helpful resources I found were Facebook groups: "Parents with PTSD/C-PTSD" at https://www.facebook.com/groups/229502014639930 and "Raising your Spirited Child" at https://www.facebook.com/groups/2348651727.

225 *tube-and-syringe-aided breastfeeding:* "Using a feeding tube at the breast with a nipple shield," LA Lactation, LLC, posted on August 15, 2020. YouTube video, https://www.youtube.com/watch?v=aB4c3xOfHHc.

225 *formula into Little T's mouth:* Thanks to supplemented feeding, Little T gained more than one ounce every day. This has made me forever grateful for the life-saving supplementation. During the national shortage, friends organized milk trains, where breastfeeding parents with an excess of milk can donate their milk to parents who rely on formula. When the bubble of global capitalism ruptures, our improvised villages must provide the mutual aid necessary to navigate crises like these. At the time of writing this book, there was a severe formula shortage due to contamination at a major US production plant. Access to safe and affordable formula is a reproductive justice, early childhood, and disability justice issue. For more, see Arohi Pathak et al., "The National Baby Formula Shortage and the Inequitable U.S. Food System," Center for American Progress, June 17, 2022, https://www.americanprogress.org/article/the-national-baby-formula-shortage-and-the-inequitable-u-s-food-system/.

225 *neonatal intensive care:* Filipina mothers have been found to have lower birth weight infants than white mothers. As cited in Joyce R. Javier, Lynne C. Huffman, and Fernando S. Mendoza, "Filipino Child Health in the United States: Do Health and Health Care Disparities Exist?," *Preventing Chronic Disease* 4, no. 2 (2007), A36.

226 *breast milk antibodies:* Angela Garbes, "The More I Learn About Breast Milk, the More Amazed I Am." *The Stranger*, August 6, 2015, https://www.thestranger.com/features/2015/08/26/22755273/the-more-i-learn-about-breast-milk-the-more-amazed-i-am.

226 *a neural brake:* breathing as a brake on the sympathetic nervous system: Porges, *The Polyvagal Theory*.

227 *felt like the gaping hole:* The somatic feeling of inadequacy is common for those who have experienced childhood neglect, abuse, and other adverse childhood experiences as well as unresolved historical trauma. See Helene Berman et al., "Laboring to Mother in the Context of Past Trauma: The Transition to Motherhood," *Qualitative Health Research* 24, no. 9 (2014), 1253–64, doi: 10.1177/1049732314521902. It's also common for those who parent with a spectrum of disabilities. See "Parenting," *Disability Visibility Project* (blog), https://disabilityvisibilityproject.com/tag/parenting/.

227 *federal policy in many other countries:* Amanda Holpuch, "The US has terrible family policies. How do expats fare abroad?" *The Guardian*, September 27, 2021, https://www.theguardian.com/money/2021/sep/27/us-terrible-family-policies-expats-netherlands-ireland-japan.

228 *soak up lochia:* Sarah Bradley, "Postpartum Bleeding: What You Need to Know About Lochia," *Parents*, last modified October 8, 2019, https://www.parents.com/pregnancy/my-body/postpartum/bleeding-after-childbirth-what-you-need-to-know-about-lochia/.

228 *"significant medical event":* Angela Garbes, *Like a Mother: A Feminist Journey through the Science and Culture of Pregnancy* (New York: Harper Wave, 2018).

228 *her delicious nilaga:* Lisa and Drew's Nilaga. Serves 4–6. Ingredients: 3 pieces beef shank, 1 tablespoon salt, 6–8 garlic cloves, ½ yellow onion, 1 tablespoon peppercorn, 2 tablespoons vegetable oil (olive oil is fine too), ½ cabbage head, 2 cups sliced potatoes (can be red or russet potatoes), 2 cups sliced carrots (carrots aren't a traditional addition, but we in Cali, so. . .). Instructions: Season beef shank with salt. In a stock pot, add oil and heat to medium high. Add the seasoned beef shank, onions, and garlic, and sauté. After browning shanks on both sides, add 6 cups of water, bring to boil, and simmer (medium to medium low) for 2 hours without a lid. Skim fat from top. Add potatoes and carrots and simmer 10–15 minutes. Add cabbage and simmer for 5 minutes. To serve: eat with jasmine rice and add patis for flavor. Enjoy the beef broth goodness! Note: if you don't have 2 hours to watch over the simmering beef shank, use a pressure cooker or an Instant Pot.

229 *social childbirth:* What Richard and Dorothy Wertz call "social childbirth" in *Lying-In: A History of Childbirth in America* is a phenomenon common to many cultures across the world, wherein family and non-family alike come to help the birthing person with household chores. Examples from other cultures include the forty-day rest period known as La Cuarenta in Mexico and the forty-day period of rest among Muslim women in accordance with Islamic beliefs. For more, see Cindy-Lee Dennis et al., "Traditional postpartum practices and rituals: a qualitative systematic review," *Women's Health* 3, no. 4 (2007), 487–502, doi:10.2217/17455057.3.4.487.

229 *the twenty-four-hour cure:* Penny Simkin, *The Birth Partner: A Complete Guide to Childbirth for Dads, Partners, Doulas, and Other Labor Companions*, 5th ed. (Beverly, MA: Harvard Common Press, 2018).

230 *help regulate mine:* Buhler-Wassmann and Hibel, "Studying caregiver-infant co-regulation in dynamic, diverse cultural contexts."

230 *sensitive to infant cries:* Marc H. Bornstein et al., "Neurobiology of culturally common maternal responses to infant cry," *Proceedings of the National Academy of Sciences* 114, no. 45 (2017), E9465–73, doi:10.1073/pnas.1712022114.

231 *TENS unit:* TENS (transcutaneous electrical nerve stimulation) units provided immediate but temporary relief from pain when I had localized flare-ups in my arms and back. For more, see "What is a TENS unit and does it work?," *Medical News Today*, November 9, 2018, https://www.medicalnewstoday.com/articles/323632.

231 *more than a hundred:* The US is a global outlier on family policy: out of 193 countries, 182 have paid sick leave, 185 have paid leave for mothers, and 108 have paid leave for fathers. According to Dr. Jody Heymann of the World Policy Analysis Center, "These are very, very widespread policies and guaranteeing three months or more is very common." As quoted in Holpuch, "The US has terrible family policies. How do expats fare abroad?"

231 *LGBTQIA+ partners of any gender:* Elizabeth Wong et al., "Comparing the

availability of paid parental leave for same-sex and different-sex couples in 34 OECD countries," *Journal of Social Policy* 49, no. 3 (2020), 525–45, doi:10.1017/S0047279419000643.

231 *in Japan:* Mandy Major, "What Postpartum Care Looks Like Around the World, and Why the U.S. Is Missing the Mark," *Healthline*, March 26, 2020, https://www.healthline.com/health/pregnancy/what-post-childbirth-care-looks-like-around-the-world-and-why-the-u-s-is-missing-the-mark.

232 *from the bottom-up:* Perry and Winfrey, *What Happened to You?*, 142.

232 *regulate their internal bodily processes*: Perry and Winfrey, *What Happened to You?*, 29.

232 *because it's cathartic:* Asmir Gračanin, Lauren M. Bylsma, and Ad J. J. M. Vingerhoets. "Is crying a self-soothing behavior?," *Frontiers in Psychology* 5 (2014), 502, doi:10.3389/fpsyg.2014.00502.

232 *Little T's own neurodivergence:* On the connections between colic and neurodivergence, see Gwen Dewar, "Colicky babies and brain chemistry: Understanding the effects of temperament and pain sensitivity," *PARENTING SCIENCE*, last modified October 2017, https://parentingscience.com/colicky-babies-and-brain-chemistry/.

232 *inherited epigenetically from me:* On the transgenerational inheritance of altered stress response, see Stephen G. Matthews and David I. Phillips, "Transgenerational inheritance of stress pathology," *Experimental Neurology* 233, no. 1 (2012), 95–101, doi:10.1016/j.expneurol.2011.01.009.

233 *"emotional health to our descendants":* Menakem, *My Grandmother's Hands*.

234 *HPA axis:* The hypothalamic-pituitary-adrenal axis is a neuroendocrine system responsible for releasing stress hormones to mobilize our bodies in response to threat. For more on the scientific role of the HPA axis in stress response, see Sean M. Smith and Wylie W. Vale, "The role of the hypothalamic-pituitary-adrenal axis in neuroendocrine responses to stress, *Dialogues in Clinical Neuroscience* 8, no. (2006), 383–95, doi: 10.31887/DCNS.2006.8.4/ssmith.

For more on HPA dysregulation and PTSD, see Sanil Rege and James Graham, "Post Traumatic Stress Disorder (PTSD) – A Primer on Neurobiology and Management," *Psych Scene Hub*, last modified August 27, 2022, https://psychscenehub.com/psychinsights/post-traumatic-stress-disorder/.

234 *somatic touch:* "What is Somatic Touch Therapy and How Can It Help?" *Somatic Healing Therapy*, https://www.somatichealingtherapy.com/resources/what-is-touch-trauma-therapy.

234 *EMDR:* For more on EMDR (eye movement desensitization and reprocessing), see "Trauma Therapy Using EMDR," Cindy Chen, PsyD, https://drcindychen.com/trauma-therapy-using-emdr.

234 *EMDR treats the flame:* quote from Brock Weedman, therapist in Bellevue, Washington.

235 *parenting classes:* The parenting classes we took were through our son's public school. They offered the Incredible Years series, which focuses on positive

reinforcement to help with the social-emotional development of children ages zero to twelve. For more, see https://incredibleyears.com/.

235 *play therapist:* Play therapy didn't work well for our child, but that may have been because of the particular practitioner we tried. For more on play therapy, see "What Is Play Therapy?," Georgia State University College of Education & Human Development, https://education.gsu.edu/cps/researchoutreach/play-therapy-training-institute/what-is-play-therapy/.

235 *occupational therapist:* Occupational therapy did wonders for identifying and helping with our child's motor-related neurological development. For more, see "Occupational Therapy," Nemours KidsHealth, January 2020, https://kidshealth.org/en/parents/occupational-therapy.html.

235 *parent-child interaction therapist:* Parent-child interaction therapy (PCIT) was a great way to have structured support for us as parents to better support our child. It is a rigorous, evidence-based protocol to help children who may exhibit so-called behavioral problems but actually have specific relational needs yet to be met. For more, see "What Is Parent-Child Interaction Therapy (PCIT)?," PCIT International, http://www.pcit.org/what-is-pcit.html.

235 *occupational therapy exercises:* Some of these exercises included swinging and rocking motions to activate his vestibular system and spending time in a small toy-circus tent to allow him to return to baseline after sensory overload.

235 *music therapy:* "The Safe and Sound Protocol (SSP)," Unyte Integrated Listening Systems, https://integratedlistening.com/ssp-safe-sound-protocol/.

235 *learned about this form of music therapy:* Porges, discussion.

235 *Little T's baseline was much calmer:* "Anxiety is a bodily state—his body wants to move. Stillness for people with trauma histories [even from generations past] is the worst possible state. The music modulates. The vocal music is altered by computer to amplify prosody and intonation shifts and shifting frequencies. We've superimposed filters to aid with breathing and circulation rhythms." Porges, discussion.

236 *criminalized abortion:* While some conservatives claim that abortion is a traumatic event, UN experts call out the criminalization of abortion as a "retrogression" that paves the way for human rights abuses. "Joint web statement by UN Human rights experts on Supreme Court decision to strike down Roe v. Wade," United Nations Office of the High Commissioner, June 24, 2022, https://www.ohchr.org/en/statements/2022/06/joint-web-statement-un-human-rights-experts-supreme-court-decision-strike-down.

Watershed

237 *this conversation is about us:* Arlene Biala, "mostly water," in *one inch punch* (Cincinnati, OH: Word Poetry, 2019), 58.

238 *"So Much Trouble In The World":* Bob Marley & The Wailers, "So Much Trouble in the World," track 1 on *Survival*, Tuff Gong, 1979.

239 *patterns from fragments:* Lauret Savoy, *Trace: Memory, History, Race, and the American Landscape* (Berkeley, CA: Counterpoint, 2016).

239 *the Pasig River:* For a history of the Pasig, see Reynaldo G. Alejandro, *Pasig: River of Life* (Philippines: Unilever, 2000).

240 *the Pasig River ferry:* "Pasig River Ferry Service," Academic Dictionaries and Encyclopedias, https://en-academic.com/dic.nsf/enwiki/10886268.

241 *borne the brunt of Manila's history:* While the Pasig bore the brunt of history, it continues to bear the burden of modern-day violence and corruption: See Howard Johnson and Virma Simonette, "Lives lived and lost along Manila's Pasig river," *BBC News*, August 6, 2019, www.bbc.com, https://www.bbc.com/news/world-asia-49203752.

242 *industrial waste:* "The Grand Canal of Pasig" in *Urban Innovations*, Asian Development Bank, September 2008, https://www.adb.org/sites/default/files/publication/27888/grand-canal-pasig.pdf.

242 *bodies also piled up:* In her essay "Mississippi Montage," Colleen McElroy writes of rivers across the world that have flowed red with blood. Of the Mississippi carrying "bodies of soldiers and slaves . . . in the current like logs" and the Danube run red from Serbian women and children shot execution-style by Germans during the war. These bodies are the most obvious sediment that builds from conquest and betrayal. From being continuously used as a crossroads and a means to an ulterior end. The Pasig is one of these rivers run red. See Colleen J. McElroy, *A Long Way from St. Louie* (Saint Paul, MN: Coffee House Press, 1997).

242 *like ballast:* Luzviminda Francisco, "The first Vietnam: The U.S.-Philippine War of 1899," *Bulletin of Concerned Asain Scholars* 5, no. 4 (1973), 2–16, doi: 10.1080/14672715.1973.10406345.

242 *swallowed all hues:* Stephanie N. Gilles, "Revitalization of the Pasig River Through the Years: Bringing a Dying Ecosystem Back to Life," https://www.academia.edu/5765195/Revitalization_of_the_Pasig_River_Through_the_Years_Bringing_a_Dying_Ecosystem_Back_to_Life.

243 *traumatic growth:* For more on post-traumatic growth, see Lorna Collier, "Growth after trauma," *American Psychological Association* 47, no. 10 (2016), 48.

243 *reclaimed waterway:* One of these visionaries who also put in the work was the late Gina Lopez: Prosy Abarquez Dela Cruz, "Gina Lopez's Legacy: Social-impact Environmentalist for Pasig River's Revival," *Asian Journal*, August 21, 2019, https://www.asianjournal.com/life-style/lifestyle-columnists/gina-lopezs-legacy-social-impact-environmentalist-for-pasig-rivers-revival/.

243 *nearly a decade after:* "Award-winning rehab of 'biologically dead' Pasig River: How it happened," *Interaksyon*, October 22, 2018, https://interaksyon.philstar.com/trends-spotlights/2018/10/22/136383/award-winning-rehab-of-biologically-dead-pasig-river-how-it-happened/.

244 *forty-seven minor tributaries:* "Pasig River Rehabilitation Program," in *2020 Annual Report*, Department of Environment and Natural Resources, https://

www.denr.gov.ph/images/DENR_Accomplishment_Report/Pasig_River_Rehabilitation_Program_opt.pdf.

244 *Pictures of before and after:* "PAGCOR funds Estero de Paco clean-up drive," Philippine Amusement and Gaming Corporation, January 21, 2013, https://www.pagcor.ph/press-releases/pagcor-funds-estero-de-paco-clean-up-drive.php.

244 *invasive weeds that fishermen:* Jonathan L. Mayuga, "DENR to tap small fishers to clear Pasig River, Laguna de Bay of water hyacinth," *BusinessMirror*, March 18, 2021, https://businessmirror.com.ph/2021/03/18/denr-to-tap-small-fishers-to-clear-pasig-river-laguna-de-bay-of-water-hyacinth/.

244 *it's not that we are polluted:* "People forget that the consequences of trauma are natural reactions—when people feel sad, depressed, hurt, those are normal reactions to traumatic events. All of these feelings are good; these are natural responses to oppression. The goal is not for us to become well-adjusted to an oppressive society; we want oppression to end." E. J. R. David, in discussion with the author, September 16, 2019.

244 *"or the rivers of the world":* June Jordan, "The Creative Spirit and Children's Literature," in *Revolutionary Mothering: Love on the Front Lines*, ed. Alexis Pauline Gumbs, China Martens, and Mi'ia Williams (Oakland, CA: PM Press, 2016).

245 *The Estero del Paco cleanup:* Anna Valermo, "Cleaning up Manila's Pasig River, one tributary at a time," *Manila Bulletin*, June 30, 2017, https://www.pressreader.com/philippines/manila-bulletin/20170630/281689729828064; and Howie Severino, "I-Witness: 'Estero City,' dokumentaryo ni Howie Severino (full episode)," GMA Public Affairs, posted on June 18, 2017, Youtube video, https://www.youtube.com/watch?v=Ao6g9tEmNy0.

245 *further filtered the water:* Melvic Cabasag and Jen Santos (Bantay Kalikasan), in discussion with the author, December 17, 2019. For more, see "About Us," *ABS-CBN Lingkod Kapamilya*, https://foundation.abs-cbn.com/bantay-kalikasan/about.

See also Brotons, Javier Coloma. *The Communication Story of Estero de Paco*. https://events.development.asia/system/files/materials/2016/12/201612-communication-story-estero-de-paco.pdf.

245 *"ecosystems and societies":* adrienne maree brown, *Emergent Strategy: Shaping Change, Changing Worlds* (Oakland, CA: AK Press, 2017), 3.

245 *further cleanup:* Eligia D. Clemente, "Evaluating the Water Quality Contribution of Estero de Paco to Pasig River and Manila Bay, Philippines," *ES3 Web of Conferences* 148, no. 3 (2020), 07010, doi:10.1051/e3sconf/202014807010.

246 *ten thousand pesos a month:* This amount was about 190 US dollars a month in 2019. This amounts to the low end of the monthly income scale in Manila. For more, see "2020 Occupational Wages Survey (OWS)," Republic of the Philippines Philippine Statistics Authority, February 2, 2022, https://psa.gov.ph/press-releases/id/165967.

247 *could no longer be cleaned:* Neil Arwin Mercado, "Duterte orders abolition of Pasig River Rehabilitation Commission," *Philippine Daily Inquirer News*,

November 14, 2019, https://newsinfo.inquirer.net/1189665/duterte-orders-disestablishment-of-pasig-river-rehab-commission.

247 *the presence of connection: The Wisdom Of Trauma*, directed by Maurizio Benazzo and Zaya Benazzo (2021; Sebastopol, CA: Science and Nonduality), https://thewisdomoftrauma.com/.

247 *"I have clean water in my heart":* Ate Angie, in discussion with the author, December 17, 2019.

248 *call centers:* "Call Center Outsourcing Philippines: A bright future awaits," *The Manila Times*, September 5, 2022, https://www.manilatimes.net/2022/09/05/public-square/call-center-outsourcing-philippines-a-bright-future-awaits/1857304.

248 *return to earth:* Alfredo Mahar Lagmay et al., "Street floods in Metro Manila and possible solutions," *Journal of Environmental Sciences* 59 (2017), 39–47, doi:10.1016/j.jes.2017.03.004.

248 *10 percent of wastewater:* Vicente B. Tuddao, "Updates on Domestic Wastewater Management in the Philippines," Water Environment Partnership in Asia, March 2021, http://wepa-db.net/pdf/meeting/20210301/10_Philippines.pdf.

248 *coordinated way:* Hyacinth Tagupa, "Beyond river clean-ups," *Philippine Daily Inquirer News*, June 18 2021, https://opinion.inquirer.net/141265/beyond-river-clean-ups.

249 *"come back to life and answer us":* Spenta Kandawalla, in discussion with the author, February 11, 2019.

249 *a source called interdependence:* Our nervous systems are models of interdependence. For more about the interdependence of our nervous systems, endocrine systems, immune systems see Katlein França and Torello M. Lotti, "Psycho-Neuro-Endocrine-Immunology: A Psychobiological Concept," in *Ultraviolet Light in Human Health, Diseases and Environment*, ed. Shamim I. Ahmad (Cham, Switzerland: Springer International, 2017), 123–34.

249 *source of the name "Pasig":* Ross Flores Del Rosario, "History of the Name Pasig," Amused of Pasig, July 4, 2012. http://amusedofpasig.blogspot.com/2012/07/mutya-ng-pasig.html.

249 *The COVID-19 pandemic:* On how the COVID-19 pandemic has shown that everything is connected: *Maliya V. Ellis and Kevin Lin, "A Harvard Epidemiologist's Reminder of Our Common Humanity," The Harvard Crimson*, April 17, 2020, https://www.thecrimson.com/article/2020/4/17/nancy-krieger-coronavirus/. On global interdependence and COVID-19: The Lancet, "COVID-19: Learning as an interdependent world," *The Lancet* 398, no. 10306 (2021), 1105, doi:10.1016/S0140-6736(21)02125-5.

249 *entrenched over time:* Doidge, *The Brain That Changes Itself.*

249 *public health crisis:* Kathryn M. Magruder, Katie A. McLaughlin, and Diane L. Elmore Borbon, "Trauma is a public health issue," *European Journal of Psychotraumatology* 8, no. 1 (2017), 1375338, doi:10.1080/20008198.2017.1375338.

249 *larger environments:* "[The tension between flexibility and entrenchment in

epigenetic changes] is the whole point. What sustains enduring change is when there aren't great systemic changes to one's environment. . . . If we corrected cycles of poverty and cycles of violence, we can ask, 'Are epigenetic changes helping us with this?' The biology doesn't remove from us the obligation of social changes; it makes it more of a mandate. Let's change those environments." Rachel Yehuda, in conversation with the author, March 1, 2021.

249 *more equitable and coordinated systems:* This approach is not the same as a resilience-based approach; it is a transformation-based approach that focuses on changes in the environment rather than solely changes within the behavior of populations. As E. J. R. David says, "The problem is not our resilience. The problem is a world that always requires us to be resilient. That's where the focus needs to be." David, discussion.

250 *collective and political:* Herman, *Trauma and Recovery*.

250 *traumatic systems of oppression:* Aurora Levins Morales writes, "Examining psychological trauma inevitably leads us to the most widespread source of trauma, which is oppression." Morales, *Medicine Stories*, 16.

250 *systemic political and cultural change:* "That's what I'm planning to do with [my research on] transgenerational epigenetic inheritance. My mother is from South America; I have a Black American father. I suffer from metabolic issues—Black people are more likely to have higher blood pressure [regardless of behavior]. These traumas live on, and we need to change science, policy, and medicine to address these issues." Bianca Jones Marlin (Columbia University), in conversation with the author, February 2, 2021.

250 *all living things:* My ecosystem view of trauma-wise care and holistic well-being grew from the work of Nancy Krieger. See Nancy Krieger, *Ecosocial Theory, Embodied Truths, and the People's Health* (New York: Oxford University Press, 2021).

251 *We have a long way to go:* For a summary of Pasig rehabilitation efforts and the long way left to go, see Jonathan L. Mayuga, "What's next? The many noble but unsuccessful attempts to rehabilitate Pasig River," *BusinessMirror*, April 4, 2018, https://businessmirror.com.ph/2018/04/04/whats-next-the-many-noble-but-unsuccessful-attempts-to-rehabilitate-pasig-river/.

251 *unmyelinated nerve:* Lakna, "Difference Between Myelinated and Unmyelinated Nerve Fibers," Pediaa.Com, September 13, 2017, https://pediaa.com/difference-between-myelinated-and-unmyelinated-nerve-fibers/.

251 *troubled rivers:* This is a nod to Komunyakaa, "Anodyne." I didn't know what the word Anodyne meant until I looked it up as I wrote this final endnote. According to Wikipedia it means: "a drug used to lessen pain through reducing the sensitivity of the brain or nervous system."

Index

NOTE: Page numbers for notes are differentiated from those for regular text by an "n" following the page number.

About the Author

JEN SORIANO (she/they) is a writer, movement builder, independent scholar, and performer. She is the author of *Making the Tongue Dry* and coeditor of *Closer to Liberation: Pin[a/x]y Activism in Theory and Practice*. Soriano's work won the *Fugue* Prose Prize and the Penelope C. Niven Prize for Creative Nonfiction and earned her fellowships from the Vermont Studio Center, Hugo House, and Jack Jones Literary Arts. She received a BA in history and science from Harvard and an MFA from the Rainier Writing Workshop at Pacific Lutheran University, and she has served as poet-in-residence for the Washington Physicians for Social Responsibility. Soriano lives in Seattle with her family and a growing community of dreamers between Lake Washington and the Salish Sea.